GEOMETRIC
TOLERANCING

A Text-Workbook

GEOMETRIC TOLERANCING

Richard S. Marrelli
Los Angeles Pierce College

GLENCOE
A Macmillan/McGraw-Hill Company
Mission Hills, California

This work is dedicated to my father,
Michael S. Marrelli
who showed me how to be a man and a father,
and to my mother,
Antoinette Guadagnolo Marrelli
who taught me about loving and caring.

Send all inquiries to:
Glencoe/McGraw-Hill
15319 Chatsworth Street
P.O. Box 9609
Mission Hills, CA 91346-9609

Library of Congress Catalog Card Number: 82-84363

ISBN 0-02-829810-1

5 6 7 8 9   93 92 91

CONTENTS

PREFACE

Not many teaching texts are commercially available on geometric tolerancing. Until the appearance of this text, works in print were either large and expensive, with many pages of seldom used applications, or they were short and simplistic—not sufficiently broad and deep to be very useful. *Geometric Tolerancing* is designed to fill the need for a text that is thorough and at the same time concise, containing all most readers need to know about the subject, plus a little more.

It is intended for students of mechanical drafting, design, engineering, manufacturing, and quality control, as well as for working professionals in these fields. It contains enough material for a full semester course, or portions of it can be used in a shorter program. Also, *Geometric Tolerancing* can be used effectively as a supplementary text in courses on mechanical drafting, machine design, manufacturing planning, or quality control. In addition, any individual needing to master the subject will find it an easy self-study text.

The text is written for use at the post-secondary-school level—community colleges, technical schools, universities, and company-sponsored training programs—but will serve the needs of advanced high school courses as well. A preliminary draft of *Geometric Tolerancing* was prepared by the author in 1980 in order to collect constructive criticism and suggestions for improvement from a sampling of students and professionals for whom the text is intended. It was used as a working text at the community college level for two years and was reviewed by teachers and by design supervisors of large and small manufacturing firms. Their comments and suggestions, as well as those of students, are reflected in the finished product.

ANSI Y14.5M-1982

Geometric Tolerancing is up-to-date, presenting the most recent changes in American geometric tolerancing practice. The text incorporates the revised standard on dimensioning and tolerancing completed in December, 1982, by the American Society of Mechanical Engineers under the auspices of the American National Standards Institute. The revised standard is designated ANSI Y14.5M-1982.

The previous revision of the ANSI standard was made in 1973, and it is expected that practices from both the old and new standards will be in concurrent use for a few years. Also, many geometric tolerance users have never adopted all of the 1973 revisions. For these reasons, this text includes an appendix explaining former practices that are still being applied to current drawings and that are also found on older drawings.

About thirty percent of the numerical values used in the examples and exercises in this text are metric. This is judged to be the best proportion for the current level of metric conversion in American industry. Where metric dimensions are used in the illustrations, this is noted in the captions. All metric units are millimeters; all conventional English units are inches.

Organization

This text is arranged in an orderly manner conducive to learning and to its use as a reference. Each topic is introduced by a definition and, wherever possible, by a comparison with a topic already learned. The instruction in each chapter starts with easily understood and commonly used concepts, then proceeds to the more difficult and less frequently encountered applications.

The book is divided into short, easily mastered chapters, giving the student the confidence of accomplishment. Much information is condensed into tables to provide easy comparison for students and quick reference for professionals. The numerous high-quality illustrations give the reader a clear picture of the written instruction, and the large page size allows for the optimum arrangement of the illustrations. The wide margins provide space for the student to write notes or to jot down questions to be brought up in class.

A workbook section, containing enough review and evaluation problems for an entire semester, is included with the text. It consists of tear-out sheets for each chapter, which include multiple-choice tests and practical exercises. In addition, there are seven comprehensive exercises that draw on material learned throughout the text and challenge the student to apply concepts in simulated industrial applications.

Teacher's Manual

A teacher's manual is available which outlines teaching strategies and tells how to make best use of the text. The manual also gives solutions to all the chapter tests and comprehensive exercises, and it includes additional topic tests and a final examination. With the manual and the text-workbook, the instructor will have a comprehensive, ready-to-use teaching package.

Acknowledgements

The author wishes to express appreciation to his colleagues and to the industrial experts who reviewed the preliminary draft.

Charles M. Duffy, Professor Emeritus, and Marion Reid, Associate Professor; both of the Industrial Education Department, Los Angeles Pierce College; Woodland Hills, California.

Russell G. Corbin, Engineering Group Manager; Fisher Body Division, General Motors Corporation; Warren, Michigan.

Lea R. Lange, Section Head, Design Superintendent, Drafting; Missile Systems Group, Hughes Aircraft Company; Canoga Park, California.

Sam Phillips, Manager, Computer Aided Design and Drafting Systems, and Mario Barbato, Supervisor of Checking; both of Teledyne Systems Company; Northridge, California.

Sterling Scott, Manager, Documentation, Design, and Graphic Services; Data Products Corporation; Woodland Hills, California.

Sam Fukashigi, Mechanical Engineer; Data Systems Division, Litton Industries; Van Nuys, California.

Jerry Ziger, Chief Checker; Ride and Show Engineering, WED Enterprises, a division of Walt Disney Productions; Glendale, California.

Thanks are also due to the following organizations for permission to use the material listed.

American Society of Mechanical Engineers and American National Standards Institute for the use of many illustrations from ANSI Y14.5-1973 and ANSI/ASME Y14.5M-1982, adapted with permission of the publisher.

General Motors Corporation for the diagram below Table B-2 in Appendix B.

National Tooling and Machining Association for the conversion tables in Appendix B.

Comments, suggestions for improvement, or criticism by teachers, students, and working professionals are invited. Please address letters to the author in care of Vocational Education Department, Glencoe Publishing Company, 17337 Ventura Boulevard, Encino, California 91316.

CHAPTER 1

INTRODUCTION TO GEOMETRIC TOLERANCE

1.1 What Is Geometric Tolerance?

This text is intended for students of drafting, design engineering, manufacturing, and quality control, as well as for working professionals in these areas. It is assumed that the reader already has an understanding of the concept of dimensional tolerance in engineering drawings. This text is meant to expand the concept of tolerance (total permissible error) to include the *geometry* of a part as well as its dimensions. In using the word *geometry* in relation to an object, two characteristics are referred to:

1. The shape (e.g., cylindrical, rectangular, triangular, and so forth).
2. The relationship of its features. A feature is any portion of an object, such as a surface, hole, or groove.

For example, a pulley may have a cylindrical hole with a keyway. This describes the shape of the opening. The required relationship may be that the keyway be centered on the axis of the hole and the hole be perpendicular to one face of the hub; also, the opposite hub face may need to be parallel with the first face. These are examples of geometric characteristics that must be specified on the drawing and correctly interpreted by manufacturing and quality control personnel.

An example will illustrate why it is necessary to control the geometry of a part as well as its dimensions. Below is one view of a spacer, dimensioned as shown. The function of this part is to space two other parts .750 apart, within ±.010.

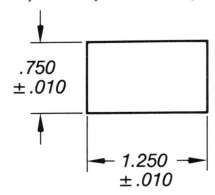

Since all the tolerances are ±.010, the part may be .010 smaller or larger in width and height. This is shown below with phantom lines. The tolerance is exaggerated for clarity.

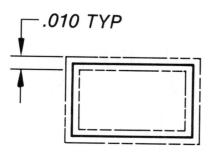

The limits of the part, therefore, can be anywhere between the phantom lines. For example, the spacer can actually be produced like this:

The part produced as shown is acceptable as far as the dimensional tolerances are concerned, but will not properly fulfill its function. What is wrong is the geometry, the shape and relationship of the surfaces. Geometry can be controlled by specifying tolerances on the geometric form as well as the linear dimensions. One way this can be done is by using local notes.

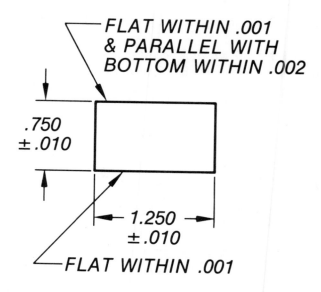

Here is the modern method, using standardized symbols.

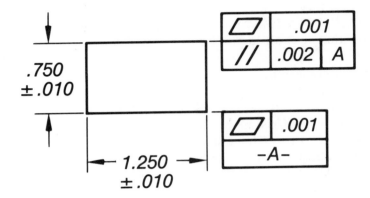

Now, regardless of the dimensional variation, the top and bottom surfaces must be flat within .001 and parallel within .002.

Geometric tolerance, then, is the permissible error in the geometric characteristics of an object—its form and the relationship of its features.

1.2 A Brief History

Until about the time of World War II, geometric form was generally not controlled on engineering drawings. It was left to the shop to see that surfaces were properly flat, circular, parallel, and so forth. However, where a high level of precision was required, this procedure was inadequate. In many cases there was an expensive number of rejects and at other times the shop was spending too many work hours obtaining very close geometric form when it was not necessary. Gradually designers began to specify the required geometric form and position tolerances by means of local notes.

The use of the symbol system known today started with the British just before World War II. Since there was much collaboration between British, Canadian, and American firms during the war, the method became known to North American engineers and was gradually adopted by North American industry. In 1956 the American Standards Association (ASA), now the American National Standards Institute (ANSI), published a standard procedure for the application of geometric form and positional tolerances. This procedure was included in its "Dimensioning and Tolerancing Standard," known as ANSI Y14.5-1956.

A *standard* is a document specifying certain methods and procedures to be used by all interested parties in a certain industry. For example, the Society of Automotive Engineers (SAE) has standards on the sizes and materials of rubber hose for automotive vehicles, and these standards are used by all the manufacturers of automobiles, trucks, and tractors.

1.3 ANSI Y14.5M–1982

The latest version of the "Dimensioning and Tolerancing Standard," adopted late in 1982, is known as ANSI Y14.5M–1982. Most manufacturing companies around the country now are using this standard or soon will start doing so. The federal government also has adopted it. Until 1966 the government had its own standard on dimensioning and tolerancing, which was known as MIL-STD-8.

The acceptance of one uniform standard in this area for the whole country was a big step forward in standardization.

ANSI Y14.5M-1982 replaces the last standard, released in late 1973 and designated ANSI Y14.5-1973. Note that the new standard has added an "M". This indicates that it is all metric—except for a few examples given in inch dimensions.

There are still some minor differences in practices among various companies, even within the same industry and the same geographic area. For instance, some companies have adopted the new ANSI symbol for diameter (∅) and use it for dimensions as well as for geometric tolerances; some companies use it only for geometric tolerances; others don't use it at all, preferring the abbreviation DIA.

All of the symbols and methods used in this text are in conformance with ANSI Y14.5M-1982. Of course, there are thousands of drawings in existence made to the 1973 and earlier standards. Appendix A provides a review of the principal differences.

CHAPTER 2

SYMBOLS

2.1 Geometric Characteristic Symbols

ANSI Y14.5M standardizes symbols for 13 geometric characteristics, in addition to four modifier symbols and several other symbols, to define exactly geometric requirements on drawings. All of these symbols are shown in Fig. 2-1 and repeated with additional data in Fig. 2-2.

All of the geometric characteristic symbols should be memorized. This is easy to do, because the shape of each symbol reminds one of the characteristic. The symbol for parallelism, for example, is two parallel lines. Perpendicularity is symbolized by two perpendicular lines. Two concentric circles are used for concentricity. The others are equally simple.

The sizes shown in Fig. 2-1 are full size, just as they are commonly used on drawings. Although ANSI Y14.5M does not specify sizes in inches or millimeters, it does standardize proportions based upon lettering height. The dimensioned sizes in Fig. 2-1 are based upon a lettering height of 5/32 (.16) inch or 4 millimeters, which is widely used. Some drafting offices using English units prefer to use 1/4 (.25) inch in place of the .24-inch dimensions given in Fig. 2-1.

Notice that all the angles are 30°, 45°, or 60°. The symbols may easily be drawn with triangles and a circle template, or with a special template made for this purpose. Templates are available with symbol sizes based upon lettering heights from 1/8 (3 mm) to 5/32 (4 mm) to 3/16 (5 mm).

The symbols are made of thick lines, drawn with the same pencil used for drawing visible lines.

The use of the symbols and their application to drawings is explained in chapters 7 through 19. Fig. 2-2 presents a quick resume of some of the data in these chapters.

2.2 Datum Identifying Symbol

The datum identifying symbol consists of a capital letter inside a box. A dash to the left and right of the letter serves to accentuate the letter.

In inches
.62 long × .31 high × .16 letter height; dashes .08 long.

In millimeters
16 long × 8 high × 4 letter height; dashes 2 long.

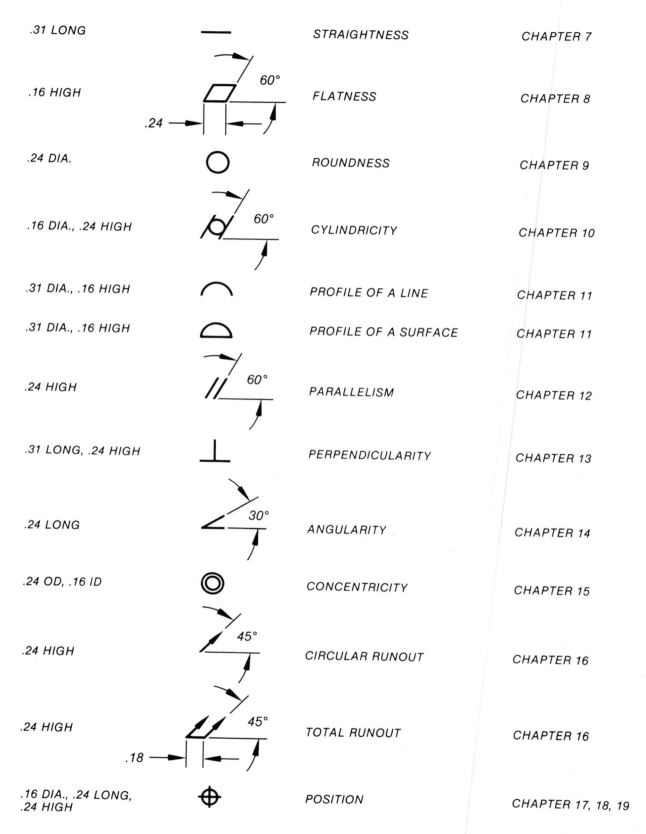

.31 LONG	STRAIGHTNESS	CHAPTER 7
.16 HIGH	FLATNESS	CHAPTER 8
.24 DIA.	ROUNDNESS	CHAPTER 9
.16 DIA., .24 HIGH	CYLINDRICITY	CHAPTER 10
.31 DIA., .16 HIGH	PROFILE OF A LINE	CHAPTER 11
.31 DIA., .16 HIGH	PROFILE OF A SURFACE	CHAPTER 11
.24 HIGH	PARALLELISM	CHAPTER 12
.31 LONG, .24 HIGH	PERPENDICULARITY	CHAPTER 13
.24 LONG	ANGULARITY	CHAPTER 14
.24 OD, .16 ID	CONCENTRICITY	CHAPTER 15
.24 HIGH	CIRCULAR RUNOUT	CHAPTER 16
.24 HIGH	TOTAL RUNOUT	CHAPTER 16
.16 DIA., .24 LONG, .24 HIGH	POSITION	CHAPTER 17, 18, 19

Fig. 2-1 Geometric characteristics shown actual size. **(a)** English units—dimensions in inches.

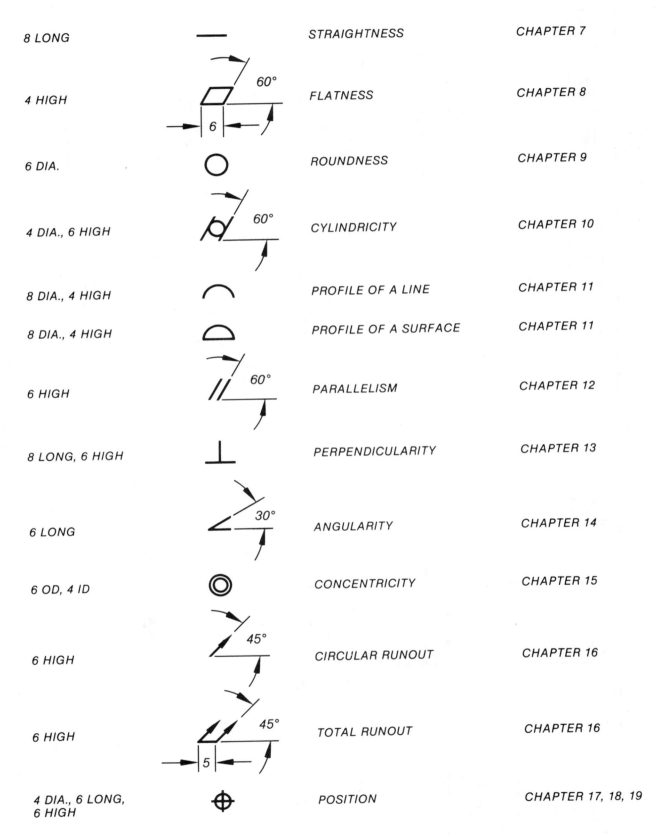

8 LONG	STRAIGHTNESS	CHAPTER 7
4 HIGH	FLATNESS	CHAPTER 8
6 DIA.	ROUNDNESS	CHAPTER 9
4 DIA., 6 HIGH	CYLINDRICITY	CHAPTER 10
8 DIA., 4 HIGH	PROFILE OF A LINE	CHAPTER 11
8 DIA., 4 HIGH	PROFILE OF A SURFACE	CHAPTER 11
6 HIGH	PARALLELISM	CHAPTER 12
8 LONG, 6 HIGH	PERPENDICULARITY	CHAPTER 13
6 LONG	ANGULARITY	CHAPTER 14
6 OD, 4 ID	CONCENTRICITY	CHAPTER 15
6 HIGH	CIRCULAR RUNOUT	CHAPTER 16
6 HIGH	TOTAL RUNOUT	CHAPTER 16
4 DIA., 6 LONG, 6 HIGH	POSITION	CHAPTER 17, 18, 19

Fig. 2-1 Geometric characteristics shown actual size.   **(b)** Metric units—dimensions in millimeters.

GEOMETRIC CHARACTERISTIC	SYMBOL	TYPE OF TOLERANCE	PERTAINS TO:	USE OF MODIFIER WITH FEATURE TOLERANCE	USE OF MODIFIER WITH DATUM
STRAIGHTNESS	—	Form tolerance	Individual feature only	Modifiers not applicable	No datum
FLATNESS	▱				
ROUNDNESS	○				
CYLINDRICITY	⌀				
PROFILE OF A LINE	⌒		Individual feature or related features		
PROFILE OF A SURFACE	⌓				
ANGULARITY	∠		Related features	Ⓢ is implied unless Ⓜ is specified	Ⓢ is implied unless Ⓜ is specified
PERPENDICULARITY	⊥				
PARALLELISM	∥				
CIRCULAR RUNOUT	↗			Ⓢ is implied unless Ⓜ is specified. If design requires that feature be Ⓜ, use positional tolerance (⊕).	Ⓢ is implied unless Ⓜ is specified. If design requires that datum be Ⓜ, use positional tolerance (⊕).
TOTAL RUNOUT	↗↗				
CONCENTRICITY	◎	Location tolerance		Ⓜ, Ⓢ or Ⓛ must be specified for each application.	
POSITION	⊕				Ⓜ, Ⓢ or Ⓛ must be specified for each application.

Fig. 2-2 Resume of data on geometric characteristics.

The meaning of *datum* is given in Section 3.5. Letters are selected in alphabetical order as required. The letters I, O, and Q are not used.

The datum surface or line is identified by placing this symbol on the drawing by any of the same means as are used for feature control frames. This will be explained in Section 2.5 and is illustrated in Fig. 2–5.

2.3 Modifier Symbols

The following symbols are used to modify geometric tolerances or the size of datums.

Ⓜ Maximum Material Condition (MMC)
　 The symbol is read "circle M."

Ⓛ Least Material Condition (LMC)
　 The symbol is read "circle L."

Ⓢ Regardless of Feature Size (RFS)
　 The symbol is read "circle S."

Ⓟ Projected Tolerance Zone (no abbreviation)
　 The symbol is read "circle P."

.24 inch DIA × .12 inch letter height
　　　　or
6 mm DIA × 3 mm letter height

These items are defined in sections 3.8, 3.9, and 3.10, respectively. The use of modifiers is explained in Chapter 6, "General Rules," and in chapters 7 through 19, as they apply.

2.4 Other Symbols

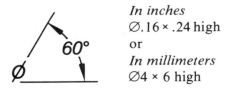

In inches
∅.16 × .24 high
or
In millimeters
∅4 × 6 high

This is the symbol for diameter. The diameter symbol replaces the abbreviation DIA on drawings, standards, and specifications. The symbol is placed *before* the dimension, as in the examples above, or *before* a diametral geometric tolerance. Henceforth, in this text, the symbol ∅ will be used to indicate diameter.

∅.16 in. or ∅4 mm

This symbol is read "all around." It is used on leaders for profile tolerances only. See Chapter 11 for its application.

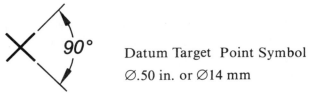

Datum Target Point Symbol
∅.50 in. or ∅14 mm

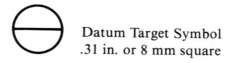

Datum Target Symbol
.31 in. or 8 mm square

Datum targets and target points are small areas or points, respectively, from which datums are located by suitable tooling on parts having rough or irregular surfaces. The use of datum targets and target points is explained in Chapter 20.

2.5 Feature Control Frame

A complete feature control frame is shown in Fig. 2–3. It is made up of a box (frame) divided into compartments containing, at the very least, a geometric characteristic symbol and a geometric tolerance value. The frame is read left to right and may contain the following:

1. Geometric characteristic
2. Tolerance
3. Tolerance modifier
4. Datum
5. Datum modifier

When required, a modifier symbol is given after the tolerance. When a datum is specified, the appropriate letter is given in the next compartment, and this may be followed by its own modifier symbol.

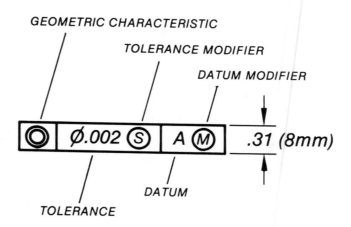

Fig. 2-3 Complete feature control frame shown actual size.

Notice that the lines are the same thickness as visible lines on a drawing. The frame is always drawn horizontally on the sheet, in the same manner as local notes.

The height of the frame is .31 (8 mm); the length varies with the requirement but is never less than 1.25 (30 mm). Fig. 2–4 shows a few typical examples in both metric and English dimensions.

The feature control frame is applied to a specific feature of the object by the manner in which it is placed on the drawing. As with dimensions, the symbol is never repeated. It is given in just one place for a particular geometric tolerance on a particular feature.

A feature may have more than one geometric tolerance. For example, on one plane surface both flatness and parallelism may be specified.

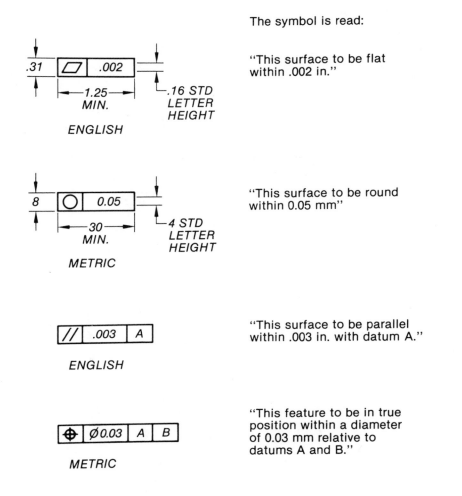

The symbol is read:

"This surface to be flat within .002 in."

"This surface to be round within 0.05 mm"

"This surface to be parallel within .003 in. with datum A."

"This feature to be in true position within a diameter of 0.03 mm relative to datums A and B."

Fig. 2-4 Examples of feature control frames.

The feature control frame is directed to the feature being controlled by one of the methods shown in Fig. 2–5.

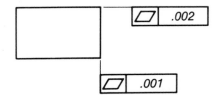

(a) Attach a side of the frame to a horizontal or vertical extension line drawn from an edge view of the feature.

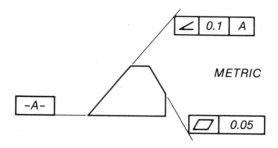

(b) Attach a corner of the frame to a slanted extension line.

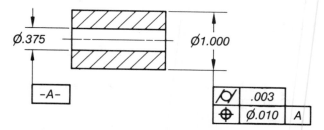

(c) Attach a side of the frame to a dimension line. This is used only when the feature is a *size* feature (a cylinder, keyway, tab, etc.).

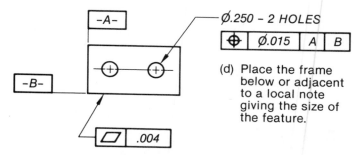

(d) Place the frame below or adjacent to a local note giving the size of the feature.

(e) Attach the left or right side of the frame at its midheight to a leader which points to the controlled feature or to an extension line drawn from it.

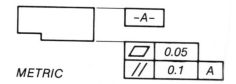

(f) Stack or hang the frame on another frame pertaining to the same feature. It is good practice to draw the frames so that the right end of a shorter frame aligns with a compartment line of a longer frame.

Fig. 2-5   Methods of applying feature control frames and datum identifying symbols to drawings.

CHAPTER 3

TERMS

In this chapter, some of the terms used in geometric tolerancing that may be unfamiliar to the reader are defined. In some cases, a definition may be somewhat specialized for this subject and different from the usual English usage. The terms are given in the approximate order in which they occur in the text.

3.1 Feature

A *feature* is any portion of an object. A feature may be a point, an edge, a center line, or a plane or curved surface. It can also be a size, such as the width of a slot, groove, or tab, or the diameter of a cylinder, in which case it is called a *size feature*.

3.2 Error

An *error* is an unintentional variation from a desired dimension or geometric form, location, or orientation. The error is acceptable and is not *wrong* unless it exceeds limits (tolerances) given on the drawing.

3.3 Element

An *element* is any line, real or imaginary, that can be drawn on a surface, including flat surfaces (planes) and curved surfaces (cylinders, cones, spheres). All the elements of a plane are straight lines in any direction. Elements of a cylinder may be circles, all with the same diameter as the cylinder (circular elements), or they may be straight lines parallel to the axis (longitudinal elements).

Elements of a cone may be circles, all of different diameters, on imaginary planes perpendicular to the axis, or they may be straight slanted lines on the surface, converging at the apex.

3.4 Radial Line

The word *radial* has the same root meaning as the words *ray* and *radiate*. It means pointing directly toward or away from a center. The sun's rays are radial lines, for example, because they extend outward from a center, the sun. A leader on a drawing is a radial line when it is directed to a circle, because it points to the center. A radial line does not have to intersect the center; it is just the direction of it that makes it radial.

3.5   Datum

A *datum* is a theoretically exact feature from which dimensions may be taken or from which the geometric form or position of another feature may be determined. The use of datums in connection with geometric tolerances is explained in chapters 4 and 20.

3.6   Tolerance Zone

The *tolerance zone* is the area taken up by the total amount of permissible error in a dimension or in a geometric form or position. The shape of the tolerance zone depends upon the nature of the geometric characteristic. It may be any of the following figures.

Two-Dimensional	Three-Dimensional
A rectangle	A rectangular solid
A circle	A cylinder
The space between two concentric circles	The space between two concentric cylinders

The tolerance zone is described for each geometric characteristic in the chapters that follow.

3.7   Basic Dimension (BSC)

A *basic dimension* (abbreviated *BSC*) is a dimension that is considered to be theoretically perfect, that is to say, without tolerance. Of course, the dimension cannot really be perfect. The permissible error is expressed not as a dimensional tolerance, but as a separate geometrical tolerance.

A dimension is designated as a basic dimension by enclosing it in a thick-line box. The height is .31 inch or 8 mm; the width is whatever is required to enclose the dimension. Following are some examples.

$\boxed{1.125}$	$\boxed{0.5}$	$\boxed{30°}$
Inches	Millimeters	Degrees

3.8   Maximum Material Condition (MMC)

Maximum material condition (MMC) is the condition in which a size feature (a hole, slot, or groove; or a shaft, pin, boss, or tab) contains the most material. For a shaft or other solid feature, it will be the *maximum* permitted diameter, the size in which the most metal is contained. For a hole or other hollow feature, the MMC is the *minimum* permitted diameter, wherein the hole is the smallest and the object, again, contains the most metal.

When maximum material condition is used as a modifier in a feature control frame, the tolerance specified applies only when the controlled feature or datum is at its maximum material condition (size). This will be explained in more depth in chapters 6 through 19.

3.9 Least Material Condition (LMC)

Least material condition (LMC) is the situation in a size feature when the feature contains the least material—the smallest shaft; the largest hole. When least material condition is used as a modifier in a feature control frame, the tolerance specified applies only when the controlled feature or datum is at its least material condition (size). Applications of the use of LMC are given in Section 17.6. The least material condition concept was introduced in ANSI Y14.5-1973; however, it was not at that time used as a modifier in the feature control frame.

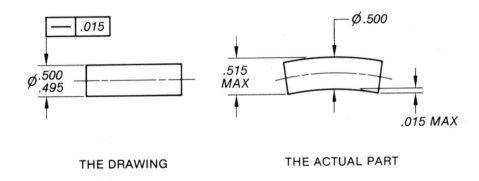

(a) Virtual Condition (Tightest Fit)

Ø.500 Max. Size Limit
+ .015 Max. Straightness Error
.515

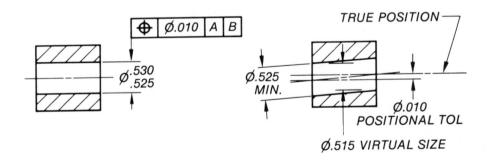

(b) Virtual Condition (Tightest Fit)

Ø.525 Min. Size Limit
− .010 Max. Positional Error
.515

Fig. 3-1 Calculations for virtual condition—a shaft and its mating part.

3.10   Regardless of Feature Size (RFS)

Regardless of feature size (RFS) means exactly what it says. When the modifier Ⓢ is used in a feature control frame, the tolerance specified applies no matter how large or small the produced size of the feature. This will be explained in more detail in chapters 6 through 19.

3.11   Virtual Condition

The *virtual condition* is the size of an object that results in the tightest fit with a mating part. This occurs with a combination of the MMC and the maximum geometric tolerance.

>   Shaft: Virtual condition = max. dia. + max. geom. tol.
>   Hole: Virtual condition = min. dia. – max. geom. tol.

Fig. 3-1 shows an example of the virtual condition of a shaft and its mating part.

3.12   Full Indicator Movement (FIM)

The *full indicator movement* (FIM) is the total movement (reading) of the pointer of a dial indicator when used to test a geometric tolerance. The dial indicator is an inspection device very commonly employed to measure geometric errors. A detailed explanation of the dial indicator is given in Chapter 5. The terms *full indicator reading* (FIR) and *total indicator reading* (TIR) were formerly used for, and had identical meaning to, FIM.

3.13   Projected Tolerance Zone

Projected tolerance zone refers to a positional tolerance zone that is moved out of the part being controlled into the mating part. The moved (projected) distance is usually equal to the thickness of the mating part and is specified by a number value followed by the symbol Ⓟ (read "circle P") in a separate frame attached below the feature control frame (see Fig. 3–2). Applications of projected tolerance zone are given in Section 18.4.

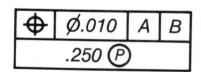

Fig. 3-2   Projected tolerance zone applied to a feature control frame.

CHAPTER 4

DATUMS

4.1 Definition

A *datum* is a theoretically exact feature from which dimensions may be taken or the geometric form or position of another feature may be determined.

4.2 Datum Plane and Datum Feature

When using the term *datum plane,* one means a theoretically exact plane that does not really exist on the part because real surfaces are imperfect. When a part is being measured, the datum plane may be considered to be the surface of the tooling, say, a surface plate, on which the part rests. This surface, though not perfect, is very accurate and is considered perfect for measurement purposes.

The term *datum feature* is used to indicate the actual surface on the part chosen as a base for measurements. Fig. 4–1 is a greatly magnified view of a datum plane. The datum feature is imperfect, because it is not perfectly flat. So the datum feature actually contacts the datum plane only at three high points. (Remember from elementary geometry that a plane may be defined by three points.) It is these three points in contact with the tooling that determine the *datum plane* from which measurements will be made.

In many situations the smoothness, flatness, or roundness of a datum feature will have to be controlled, as other features are controlled, by specifying surface finish and geometric form tolerances.

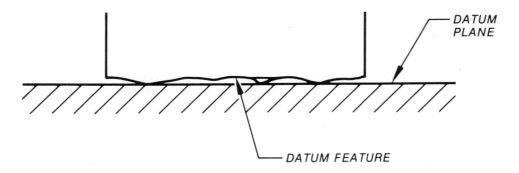

DATUM PLANE

DATUM FEATURE

Fig. 4-1 Datum plane and datum feature.

In cases where surfaces are particularly uneven, such as castings and forgings, or where the part is so thin as to be bowed or warped, it is necessary to specify the desired points of contact between the datum feature and its datum plane. ANSI Y14.5M provides a special procedure for this. It is called the datum frame concept, and it is explained in Chapter 20.

4.3 Datum Identifying Symbol

The datum identifying symbol was introduced in Section 2.2. The symbol is placed on the drawing by any of the same means as used for feature control frames (Fig. 2–5). Fig. 4–2 is a section of Fig. 2–5, showing datum identifying symbols applied to a drawing. The letter used in the box is the next available letter of the alphabet. The first datum identified on a drawing is $\boxed{-A-}$, the next one is $\boxed{-B-}$, and so forth.

Remember that the letters I, O, and Q are not used (Section 2.2).

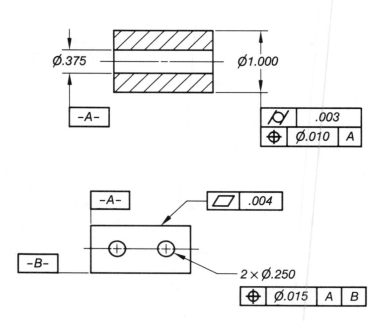

Fig. 4-2 Datum identifying symbols applied to drawings.

4.4 Selection of Datums

In selecting a feature as a datum, the following rules should be observed.

1. Select datums that are functional features. Example: A cylindrical surface that supports the part in a bearing is functional and should be used as a datum. Lathe centers, however, are not usable as datums because they do not function in the application of cylindrical parts.

2. Select corresponding features on mating parts (features that fit together).

3. Select features that are readily accessible for manufacturing and inspection.

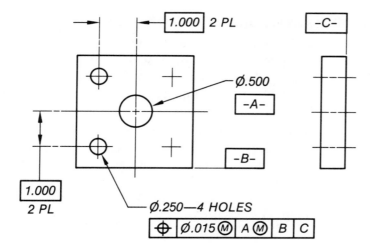

Fig. 4-3 Examples of first, second, and third datums.

4.5 First, Second, and Third Datums

For most geometric tolerances, only one datum will be required, but situations arise, chiefly with positional tolerances, that require two or three datums. In these cases the datums are given in the feature control frame in the order of their importance, and they are called the first, second, and third datums. The first datum will usually be the feature that determines the fit of the part. The second and third may also affect the fit, but to a lesser degree. An example is shown in Fig. 4–3. The central hole is selected as the first datum (A) because it fits over a mating part. The second and third datums (B and C) provide the correct angular orientation and control the squareness of the four-hole pattern.

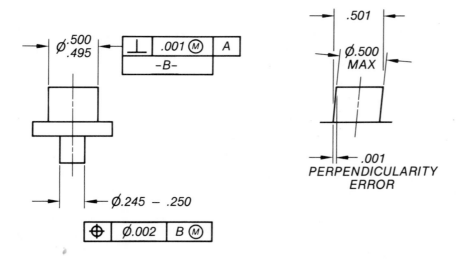

Fig. 4-4 Example of a size-feature datum specified at MMC.

4.6 Size Features as Datums

When a feature being used as a datum is not a single point, line, or plane, but has size (a size feature), complications arise because of the dimensional tolerance of the size. This variation in the size must be considered and a decision made whether the datum is to apply at the maximum material condition (Ⓜ) or regardless of feature size (Ⓢ).

An interesting situation develops when a datum feature of size is specified to apply at MMC: The produced size can actually exceed its maximum material condition. An example is shown in Fig. 4–4. Because of variation in both the diameter size and out-of-perpendicularity, the maximum effective size of Datum B is not .500 (the MMC) but .500 + .001, or .501. This is the virtual condition (size) of the feature (Section 3.11).

CHAPTER 5

THE DIAL INDICATOR

5.1 What It Is

The dial indicator is a device used to detect small movements, as small as a ten-thousandth of an inch or a thousandth of a millimeter. It consists of a glass-covered dial about the size of a small pocket watch, and a pointer that revolves over the dial. Attached to the pointer with a mechanical linkage is a spring-loaded stem or probe that extends from the side or back of the case. Very small movements of the probe are registered by the pointer and can be read on the dial.

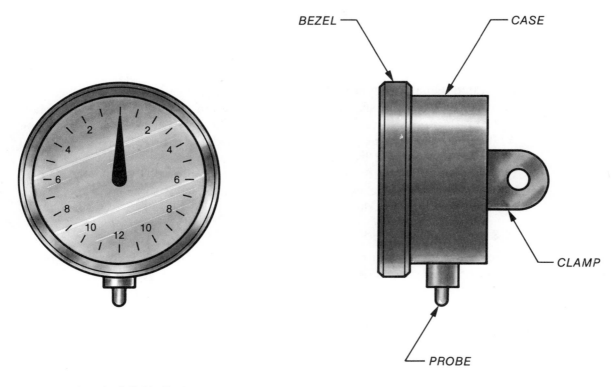

Fig. 5-1 A typical dial indicator.

A typical dial indicator is illustrated in Fig. 5–1. The pointer of this indicator will rotate 180° for a movement of the probe of .012 inch up or down. The complete movement of the probe is about .100 inch in several revolutions of the pointer. The bezel (outside rim) is attached to the dial and can be rotated by hand to align zero on the dial with the pointer wherever it happens to be when the probe is pressed against the surface to be tested. Zero does not have to be at the "12 o'clock" position.

5.2   How It Is Used

In a typical application—for example, testing parallelism—the part to be tested is placed on a surface plate and the dial is clamped to a support that is free to slide on the surface plate alongside the part (see Fig. 5–2).

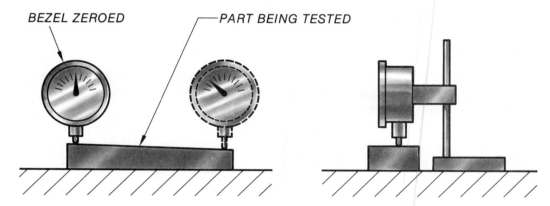

Fig. 5-2   A surface plate setup for testing parallelism.

The dial indicator is positioned above one end of the surface to be tested and lowered on its support until the probe is in contact with the surface. The probe will move upward a small amount, causing rotation of the pointer. Now the bezel is rotated by hand to "zero" the dial. To test for parallelism the dial indicator is moved across the test surface by sliding the base of the support over the surface plate. If the upper surface is not perfectly parallel with the lower surface resting on the surface plate, the probe will move up or down from its starting point and this will be registered by rotation of the pointer. The total movement of the pointer in both directions (full indicator movement, or FIM) is the parallelism error of the two surfaces. In the setup shown in Fig. 5–2, the full indicator movement is .003 and this is equal to the parallelism error.

5.3   When It Is Used

The dial indicator can be used for measuring variation in all of the geometric characteristics. A dial indicator is shown in Fig. 9–3 in a setup used for testing roundness. Figs. 16–3, 16–4, 16–6, and 16–7 illustrate applications for testing runout of curved, conical, and flat surfaces relative to one or more datums.

Testing with a dial indicator is time consuming (expensive) and therefore used only for low-production parts. Mass produced items are inspected using specially designed gages which simulate the fit of the mating part and require little or no set-up time. The part is accepted or rejected depending upon whether it fits the gage properly.

CHAPTER 6

GENERAL RULES

6.1 Introduction

There are five general rules that automatically apply to all drawings using geometric tolerances unless otherwise specified in the feature control frame, the datum identifying symbol, or in a note. The rules are numbered in the approximate order of the frequency of their use, not in the order of their importance.

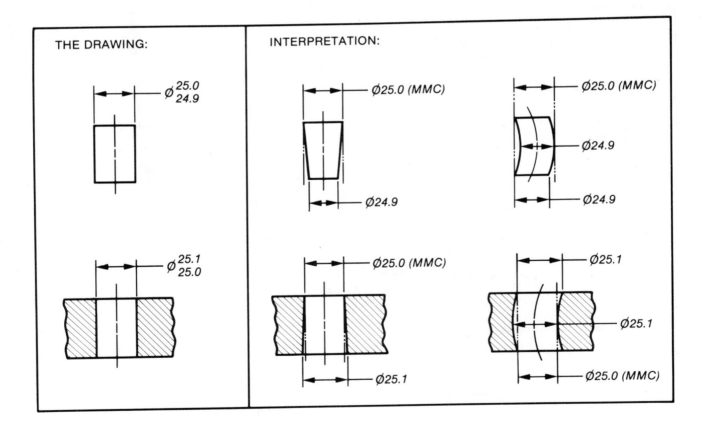

Fig. 6-1 How Rule 1 applies to an external and an internal feature. (ANSI Y14.5)

6.2 Rule 1

When no geometric tolerance is specified on an *individual* size feature, the dimensional tolerance controls the geometric form of the feature as well as the size (see Fig. 6–1).

Whether or not a geometric tolerance is given, no element of the feature shall extend beyond the MMC boundary.

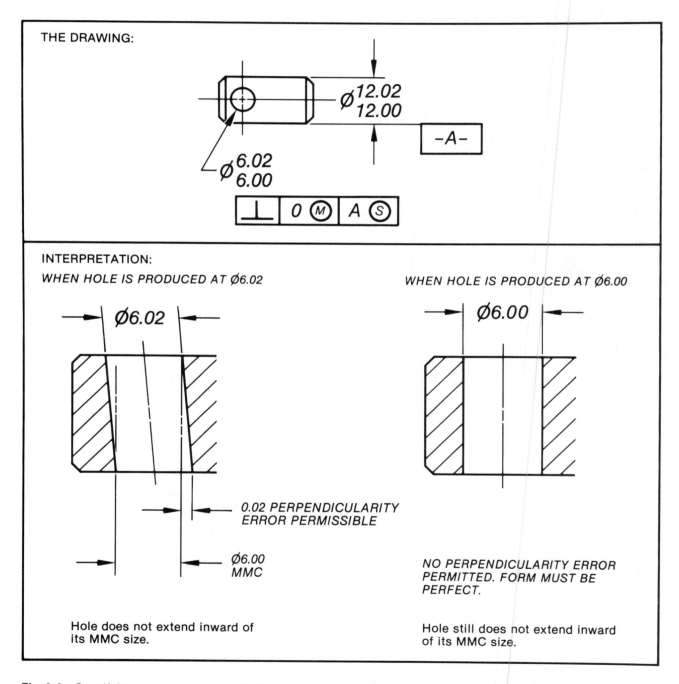

THE DRAWING:

Ø 12.02 / 12.00

−A−

Ø 6.02 / 6.00

⊥ | 0 Ⓜ | A Ⓢ

INTERPRETATION:

WHEN HOLE IS PRODUCED AT Ø6.02

WHEN HOLE IS PRODUCED AT Ø6.00

Ø6.02

Ø6.00

0.02 PERPENDICULARITY ERROR PERMISSIBLE

Ø6.00 MMC

NO PERPENDICULARITY ERROR PERMITTED. FORM MUST BE PERFECT.

Hole does not extend inward of its MMC size.

Hole still does not extend inward of its MMC size.

Fig. 6-2 Specifying zero tolerance at MMC on an interrelated feature to prevent exceeding MMC boundary.

This means that if the entire feature is at the MMC size, the geometric form must be *perfect* (no error allowed in straightness, flatness, parallelism, and so forth). An example of this is shown in the interpretation portion of Fig. 6–2.

Note that this applies only to *individual* size features, such as a single cylindrical surface or two parallel surfaces.

Exceptions to Rule 1

(a) Rule 1 does not apply when there are two or more interrelated features, such as a size feature and its datum. This occurs with the following characteristics:

    Perpendicularity

    Angularity

    Concentricity

    Runout

    Positional tolerance

    Profile tolerances—when used with a datum

Note: Parallelism is not included because the two surfaces are considered as an individual feature rather than two interrelated features.

When it is required by the design that an *interrelated* feature not exceed its MMC boundary, this can be done by specifying a zero geometric tolerance at MMC for that feature. Fig. 6–2 is an example. Zero geometric tolerance in connection with positional tolerance is discussed in Chapter 18.

(b) Rule 1 does not apply to stock items such as bars, sheets, and tubing, where established standards prescribe straightness, flatness, and so forth.

6.3   Rule 2

For *positional tolerance,* RFS, MMC, or LMC must be specified on the drawing as applicable for the feature being controlled or for its datum or for both. Formerly, all positional tolerances were *assumed* to apply at MMC when no modifier was given, but according to ANSI Y14.5M-1982 the desired modifier— MMC, LMC, or RFS—must be specified in every case.

6.4   Rule 3

For all geometric tolerances *except* positional tolerance, RFS automatically applies, both to the controlled feature and the datum. RFS does not need to be specified. If MMC is required, it must be specified.

6.5   Rule 4

Screw Threads

All geometric tolerances specified for screw threads apply to the *pitch diameter.*

When a design requirement makes it necessary to apply the tolerance to the major or minor diameter, this must be specified by a note "MAJOR DIA" or "MINOR DIA" placed below the feature control frame or datum identifying symbol, as applicable.

Gears and Splines

All geometric tolerances specified for gears and splines must specify the particular feature of the gear or spline to which they apply. This is done by lettering the name of the designated feature below the feature control frame or datum identifying symbol, as applicable. For example, specify "MAJOR DIA" or "PITCH DIA."

6.6 Rule 5

A size feature that is used as a datum and is controlled by its own geometric tolerance applies at its *virtual* condition, even though MMC is specified.

The part shown in Fig. 6–3 is an example. The equally spaced holes must be in true position relative to Datum A within .010 diameter. However, Datum A is a size feature and is controlled by its own geometric tolerance. (It must be in true position at MMC with Datum C within .003 diameter.) Although MMC is specified for Datum A, the positional tolerance of the equally spaced holes applies with A at its *virtual* size. Since it is an inside diameter, the virtual size is the smallest it can be considering its size tolerance and its positional tolerance.

Where it is not intended for the virtual condition to apply, a *zero* tolerance at MMC should be specified to control the datum feature. This has the effect of making the virtual size the same as the MMC size. In Fig. 6–3, if the positional tolerance (coaxiality) of Datum A relative to Datum C were zero, the virtual size of A (the size producing the tightest fit) would be $\varnothing 1.000$, which is the same as its MMC.

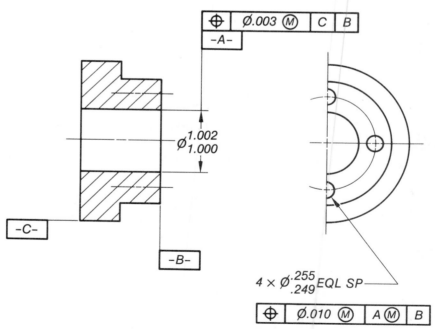

Datum A, a size feature, applies at its virtual condition: MMC ($\varnothing 1.000$) minus positional tolerance ($\varnothing .003$) equals $\varnothing .997$.

When A is produced larger than its MMC, additional tolerance is allowed in the position of the $\varnothing \frac{.255}{.249}$ holes.

Fig. 6-3 Example of the application of Rule 5.

CHAPTER 7

STRAIGHTNESS

7.1 Definition

Straightness error is the measure of how much each element in a surface or the axis of an object deviates from being a straight line. Straightness can be applied to a single surface, such as a plane or cylindrical surface, or it can be applied to a size feature, such as the diameter of a cylinder, in which case the effect is quite different, as shall be shown.

7.2 Element Straightness—Plane Surfaces

When the straightness of the elements of a flat surface is to be controlled, the feature control frame is applied to an edge view that shows the elements as a straight line, as in Fig. 7–1.

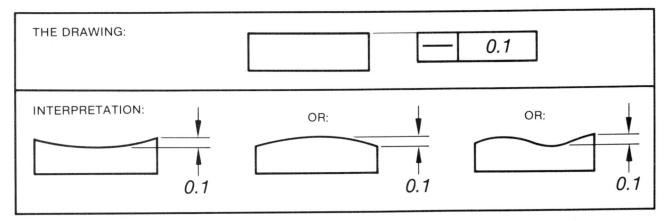

Fig. 7-1 Application of element straightness to a plane surface.

The tolerance zone is the space between two parallel straight lines.

The straightness tolerance applies to *all* the elements in the surface in the direction indicated by the placement of the feature control frame in the appropriate view. The straightness tolerance applies only in the direction shown in that view. Fig. 7–1 illustrates this. A straightness tolerance can be applied in two directions by placing feature control frames in two adjacent views, as shown in Fig. 7-2. However, if the tolerance is the same, this is identical to a flatness requirement (Chapter 8) and it is better to specify flatness.

Element straightness of a plane surface applies RFS (Rule 3, Section 6.4). Also, the straightness tolerance must be within the MMC boundary of the part (Rule 1, Section 6.2).

Fig. 7-2   Application of straightness in two directions.

7.3   Element Straightness—Cylindrical Surfaces

To control the longitudinal elements on a surface of a cylinder, the feature control symbol is applied to the straight line representing those elements or to its extension line, as in Fig. 7–3, but not to the diameter or its dimension line. A straightness tolerance thus applied controls all the longitudinal elements of the surface of the cylinder but does not control the axis.

The tolerance zone is the same as for plane surfaces. It is the space between two parallel straight lines.

The out-of-straightness of an object may occur as shown in Fig. 7–3 (d), where all the elements are bowed in the same direction. Obviously, the axis must also be bowed, but the control is not on the axis—only on the surface elements.

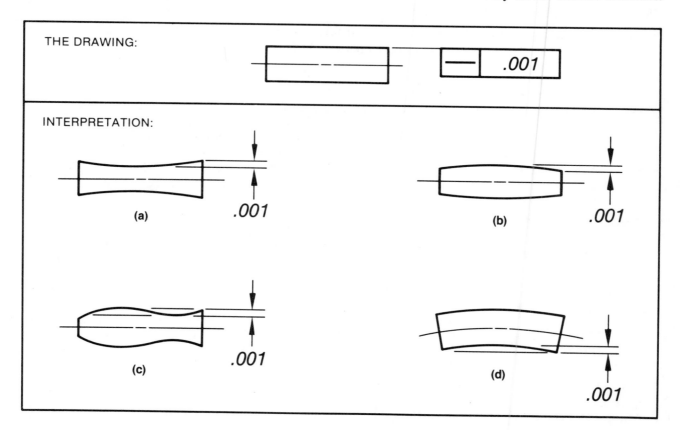

Fig. 7-3   Application of element straightness to a cylinder.

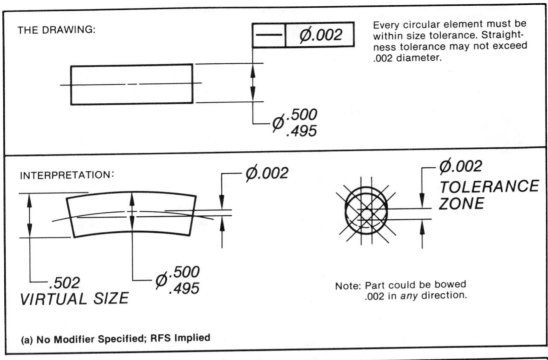

THE DRAWING:

— | ∅.002

∅ .500
  .495

Every circular element must be within size tolerance. Straightness tolerance may not exceed .002 diameter.

INTERPRETATION:

∅.002

.502
VIRTUAL SIZE

∅ .500
  .495

∅.002
TOLERANCE ZONE

Note: Part could be bowed .002 in *any* direction.

(a) No Modifier Specified; RFS Implied

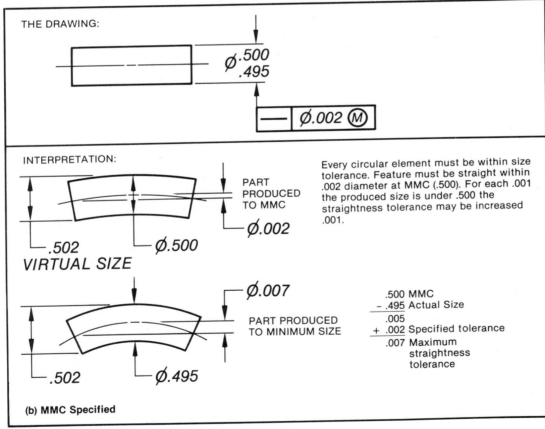

THE DRAWING:

∅ .500
  .495

— | ∅.002 Ⓜ

INTERPRETATION:

.502
VIRTUAL SIZE

∅.500

∅.002

PART PRODUCED TO MMC

Every circular element must be within size tolerance. Feature must be straight within .002 diameter at MMC (.500). For each .001 the produced size is under .500 the straightness tolerance may be increased .001.

∅.007

PART PRODUCED TO MINIMUM SIZE

.502

∅.495

.500 MMC
− .495 Actual Size
─────
.005
+ .002 Specified tolerance
─────
.007 Maximum straightness tolerance

(b) MMC Specified

Fig. 7-4 Straightness applied to a size feature.

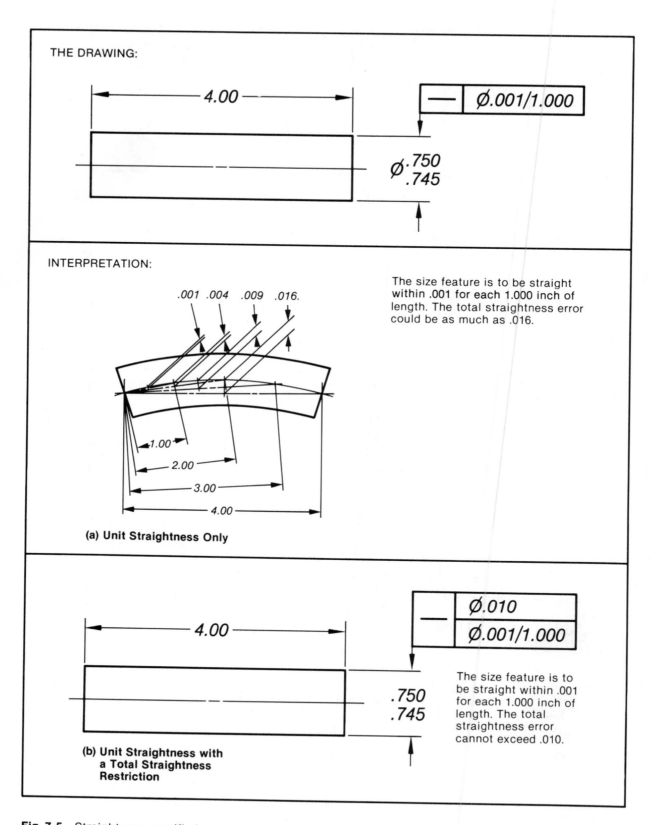

THE DRAWING:

4.00

⌀ .001/1.000

⌀ .750
.745

INTERPRETATION:

.001 .004 .009 .016.

The size feature is to be straight within .001 for each 1.000 inch of length. The total straightness error could be as much as .016.

1.00
2.00
3.00
4.00

(a) Unit Straightness Only

⌀ .010
⌀ .001/1.000

4.00

.750
.745

The size feature is to be straight within .001 for each 1.000 inch of length. The total straightness error cannot exceed .010.

(b) Unit Straightness with a Total Straightness Restriction

Fig. 7-5 Straightness specified on a unit basis.

Element straightness of a cylinder applies RFS (Rule 3, Section 6.4). Also, the straightness tolerance must be within the MMC boundary of the part (Rule 1, Section 6.2).

All of this applies equally well to objects called *disks,* which are cylinders with very short longitudinal elements.

The surfaces of other shapes that are symmetrical about their axes, such as cones, square and hexagonal bars, and so forth, are treated the same way as cylindrical surfaces.

7.4 Straightness of Size Features

It is sometimes necessary to control the straightness of a whole size feature rather than just its surface elements. This is done on the drawing by applying the straightness symbol to the size dimension or its dimension line. Two examples are shown in Fig. 7–4. An application of part (a) of the figure might be a long drive shaft that fits into bearings only at the ends. The purpose of the straightness tolerance is to limit out-of-balance due to bowing of the shaft. When a shaft or pin is to fit in a hole for its entire length, MMC may be specified as shown in part (b). This ensures that the part will fit, while allowing additional straightness tolerance if the part is not produced at its MMC.

In size-feature straightness for cylinders and cones, the tolerance zone is specified as a diameter, since the tolerance zone is the space between two straight parallel lines in all directions. This is illustrated in Fig. 7–4 (a). The diameter symbol ($\emptyset$) is placed *before* the tolerance, as shown in the figure.

Straightness of a size feature is not governed by Rule 1 (Section 6.2), because the straightness tolerance and the size tolerance are interrelated. Therefore, the boundary of the size feature *can* extend beyond the MMC, as is shown in Fig. 7–4.

7.5 Unit Straightness

In some instances, it is desirable to specify straightness on a unit basis. For example, the straightness tolerance can be specified for, say, one inch of length or three inches or 50 mm or 100 mm, and so forth. An example might be an especially long flat surface or a slender rod. The tolerance might be specified as .0005/1.00, which means that the tolerance is .0005 for each one inch of length, or .01/25 (.01 mm for each 25 mm). (See Fig. 7–5.) The reason for specifying straightness on a unit basis is to prevent the possibility of all the straightness error occurring in one area. Straightness of cold-drawn shafting from the mill, for example, is specified in thousandths of an inch per foot.

When specifying unit straightness, care must be taken that the *total* straightness error is not excessive. A straightness tolerance of .001/1.00 results in .004 in two inches, .009 in three inches, and so forth. A 20-inch feature on this basis could be .400 out of straightness. This is because the chord height of an arc is proportional to the *square* of the chord, which is illustrated in Figure 7–5 (a).

It is often necessary, therefore, to specify a *total* straightness in addition to the straightness per specified length. This is done as shown in Fig. 7–5 (b).

CHAPTER 8

FLATNESS

8.1 Definition

Flatness error is the measure of how much a plane surface deviates from being a true plane. Whereas straightness affects the elements of a surface in one direction, flatness affects all the elements in all directions. A surface can be straight in one direction but not be flat. Fig. 8–1 is an example. Notice that all the elements in the long direction (1, 2, 3, 4) are straight, but the elements in the short direction (a, b, c, d) are all curved. The surface is not flat.

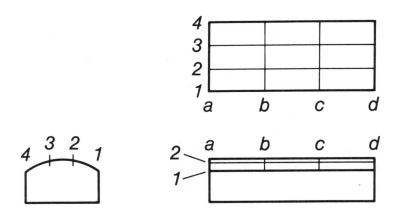

Fig. 8-1 A surface may be straight in one direction and not be flat.

A flatness tolerance zone is the space between two parallel planes (see Fig. 8–2).

8.2 Specifying Flatness on Drawings

Flatness is specified by directing the feature control frame to the edge view of the surface as shown in Fig. 8–2. The entire surface being controlled must be within the size tolerance, and no element of the surface may extend beyond the MMC boundary of the part (Rule 1, Section 6.2).

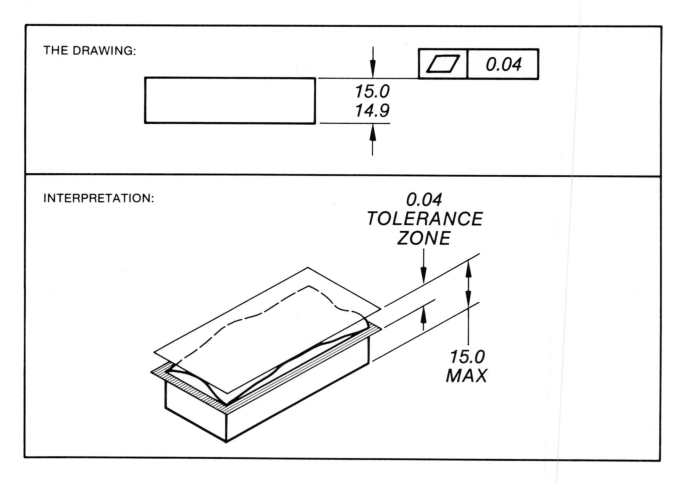

Fig. 8-2   Specification of flatness and the shape of the tolerance zone (metric).

8.3   Unit Flatness

Just as with straightness, it is sometimes desirable to specify flatness on a unit basis. The flatness tolerance might apply for, say, one inch, three inches, or 50 mm, and so forth. This is to prevent the possibility of all the flatness error occurring in one area. Fig. 8-3 shows examples of unit flatness. To avoid a buildup of excessive flatness error in the entire length, a total flatness can also be specified, as shown in Fig. 8-3 (b).

8.4   Flatness in a Specified Area

Situations occur when it is desirable to specify flatness not for an entire surface but for only a specific area. An application of this might be on a large cast or forged surface where it might be necessary to machine only a small area for mounting another part. In this case the area to be controlled is surrounded by a chain line, which is similar to a center line except that the dashes are more closely spaced. The enclosed area is then section lined. The flatness specification is applied as shown in Fig. 8-4.

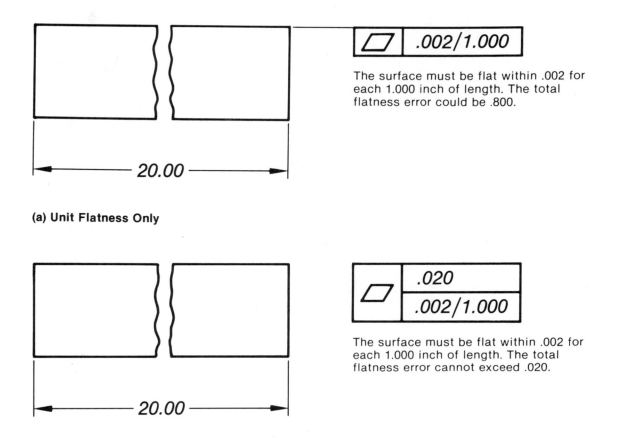

The surface must be flat within .002 for each 1.000 inch of length. The total flatness error could be .800.

(a) Unit Flatness Only

The surface must be flat within .002 for each 1.000 inch of length. The total flatness error cannot exceed .020.

(b) Unit Flatness with a Total Flatness Restriction

Fig. 8-3   Flatness specified on a unit basis.

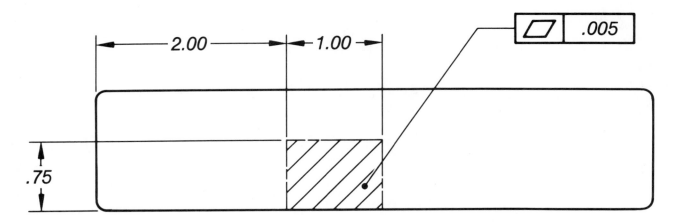

Fig. 8-4   Flatness applied to a specified area.

CHAPTER 9

ROUNDNESS

9.1 Definition

A *circle* is a curved line in which every point is equidistant from a point. This point is called the center. *Roundness error* is the measure of how much a circle deviates from having all its points equidistant from the center. It is measured as the radial space between two perfect concentric circles within which every point on the circle being tested must lie. This is shown for an exaggerated imperfect circle in Fig. 9–1. Note that the center of the two perfect circles is not necessarily the same as the theoretical center of the circle, the center shown on the print. In inspecting for roundness, neither center is ever actually located.

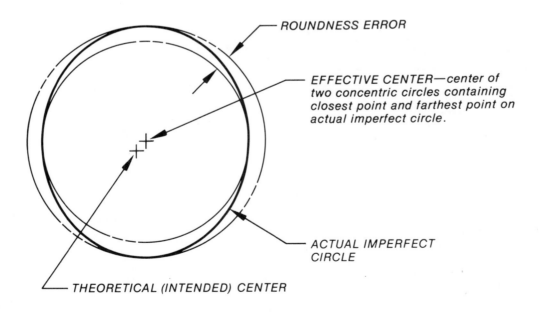

— ROUNDNESS ERROR

— EFFECTIVE CENTER—center of two concentric circles containing closest point and farthest point on actual imperfect circle.

— ACTUAL IMPERFECT CIRCLE

— THEORETICAL (INTENDED) CENTER

Fig. 9-1 Roundness error of a circle.

9.2   Roundness Tolerance

Roundness error is measured in a plane perpendicular to the axis of a circular feature, which may be a cylinder, or it may be a cone or some other surface which is circular in cross section but which varies in diameter (see Fig. 9–2). In specifying roundness on any circular surface, the roundness error of every circular element in the surface is being controlled; that is, at every cross section.

The tolerance zone is the radial space between two concentric circles. It is not a diameter, nor is it a radius, since it is not measured from the center. A roundness tolerance can be thought of as a straightness tolerance curled into a circle. Whereas straightness affects only straight elements, roundness affects only circular elements.

The roundness tolerance must be within the dimensional size limits (Rule 1, Section 6.2), and it always applies RFS (Rule 3, Section 6.4).

9.3   Specifying Roundness on Drawings

Roundness is specified on drawings as shown in Fig. 9–2. For a cylinder, the feature control symbol may be directed with a leader to the circular view rather than to the rectangular view when that is more convenient—but the symbol is never directed to both the circular view and the rectangular view.

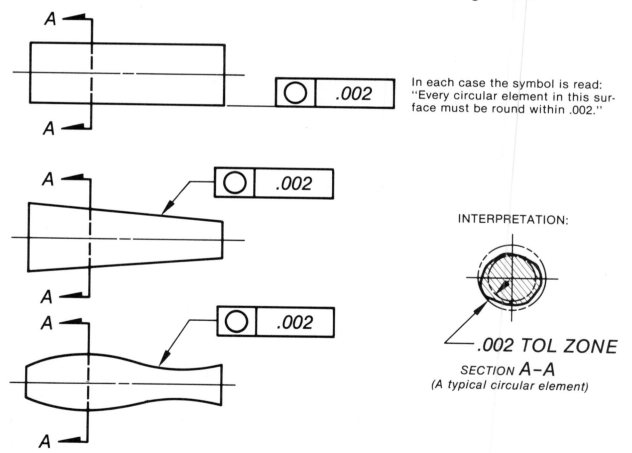

In each case the symbol is read: "Every circular element in this surface must be round within .002."

INTERPRETATION:

.002 TOL ZONE

SECTION A–A
(A typical circular element)

Fig. 9-2   Application of roundness tolerance to circular surfaces.

9.4 Measuring Roundness Error

Roundness cannot be measured directly, for to do so would require getting inside the part. Special inspection equipment that tests the part from the outside is available. Typically, the part is centered on a turntable; a dial indicator or an electronic sensor is placed in contact with the surface and the table is rotated (see Fig. 9–3).

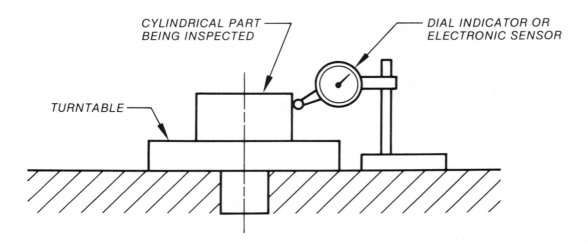

CYLINDRICAL PART BEING INSPECTED

DIAL INDICATOR OR ELECTRONIC SENSOR

TURNTABLE

Fig. 9-3 Setup for inspecting roundness.

The position of the part on the table is adjusted until the smallest readings are obtained from the indicator, when the effective axis of the part is aligned with the axis of the turntable. The turntable is then rotated to test the roundness of the part. The indicator readings every 30° or so are plotted on a large-scale circular graph until the part has been rotated 360°. The graph thus obtained is a profile of the particular circular element being tested (Fig. 9–4). A transparent film overlay of concentric circles of known distance apart at the same enlarged scale is placed over the graph. The roundness error is the distance between any two concentric circles that contain between them all the points of the profile (see Fig. 9–4). This tests the roundness of just one circular element. A sufficient number of other elements are then tested to obtain a clear idea of the roundness of all the circular elements of the surface.

Other methods of testing for roundness include revolving the part on lathe centers, on an arbor, or in a chuck, and taking readings on a dial indicator without the use of a graph. However, these are not as accurate because the dial indicator reads eccentricity between the tooling and the part and bowing of the part, in addition to roundness error.

These methods are acceptable if the FIM as tested is within the specified roundness error. But if the FIM is greater, it is not known how much of the indicator movement represents roundness error. Testing in vee-blocks is not recommended because the angle of the vee and the imperfect profile of the feature can combine to reduce or increase the apparent roundness error.

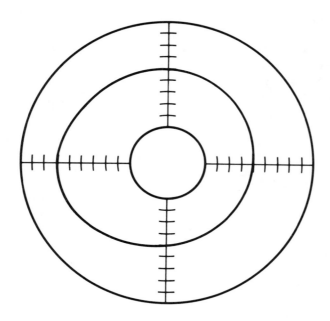
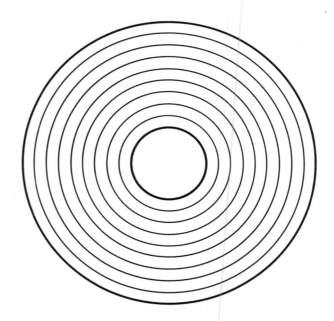

Circular Graph with Profile of Circular Element Plotted

Each graduation represents .001 inch.

Transparent Overlay

The radial distance between concentric circles represents .001 inch.

If the transparent overlay were placed over the graph, we would see that the roundness error of this circular element is .002.

Fig. 9-4   Circular graph and transparent overlay at same enlarged scale.

CHAPTER 10

CYLINDRICITY

10.1 Definition

Cylindricity is a combination of roundness and straightness. A perfect cylinder has every circular element perfectly round and every straight element perfectly parallel to the axis. *Cylindricity error* is any deviation in an actual part from this perfection.

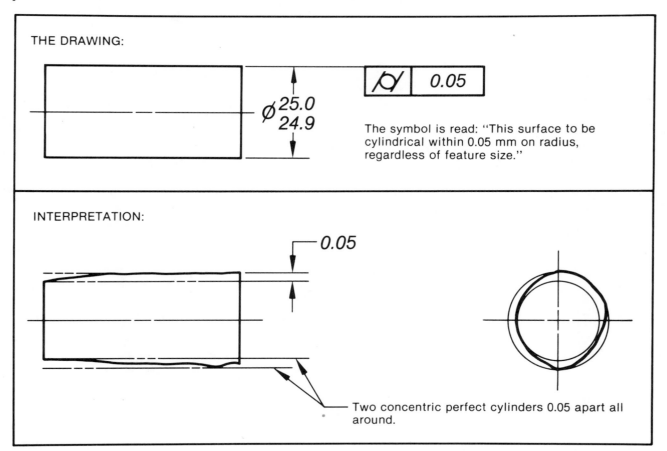

THE DRAWING:

$\phi \begin{smallmatrix} 25.0 \\ 24.9 \end{smallmatrix}$

| ⌭ | 0.05 |

The symbol is read: "This surface to be cylindrical within 0.05 mm on radius, regardless of feature size."

INTERPRETATION:

0.05

Two concentric perfect cylinders 0.05 apart all around.

Fig. 10-1 Specifying cylindricity (metric).

10.2 Cylindricity Tolerance Zone

The tolerance zone is the radial space between two concentric perfect cylinders within which every straight element and every circular element must lie (see Fig. 10–1).

A cylindricity tolerance can be thought of as a flatness tolerance curled into the shape of a cylinder. Just as flatness controls all the straight elements of a plane surface in both directions, cylindricity controls all the straight elements of a cylindrical surface in the direction of the axis and all the circular elements in a perpendicular direction.

Cylindricity tolerance must be within the dimensional size limits (Rule 1, Section 6.2) and it always applies RFS (Rule 3, Section 6.4).

10.3 Specifying Cylindricity on Drawings

Cylindricity is usually specified as shown in Fig. 10–1. An alternate method is to direct a leader from the mid-height of the feature control frame to the cylindrical surface, either in the circular view or the rectangular view.

10.4 Measuring Cylindricity Error

Cylindricity is inspected by the same methods used for roundness, primarily the circular graph method (Section 9.4). However, the longitudinal elements also must be tested for straightness and must be within the same tolerance.

CHAPTER 11

PROFILE TOLERANCES

11.1 Definition

A *profile* is the outline or contour of an object. The types of profile that will be looked at here are those which cannot be controlled by straightness, flatness, roundness, or cylindricity tolerances. They may be combinations of arcs or they

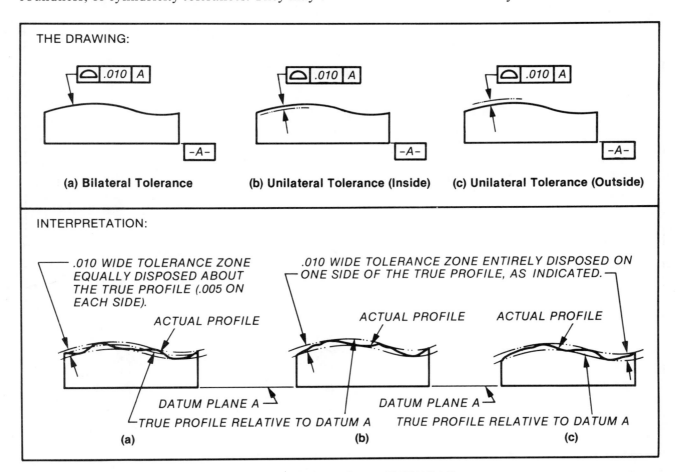

Fig. 11-1 Specifying profile tolerance. Surface tolerance shown. (ANSI Y14.5)

may be irregular curves such as those drawn with a drafter's French curve. *Profile error* is any deviation of an actual part from the desired profile.

A profile tolerance may control just the elements of a curved surface in a certain direction, as straightness does for elements on a flat or cylindrical surface. This is *profile of a line* tolerance.

Another type of profile tolerance is required when the entire surface must be controlled in both directions. This is *profile of a surface* tolerance. It does for curved surfaces what flatness does for flat surfaces and cylindricity does for cylindrical surfaces.

11.2   Profile Tolerance Zone

For profile of a line tolerance, the tolerance zone is the space between two imaginary curved lines perfectly parallel to an element of the desired profile. For profile of a surface it is the space between two imaginary curved surfaces perfectly parallel to the surface of the desired profile. This is illustrated in Fig. 11–1, in the interpretation section.

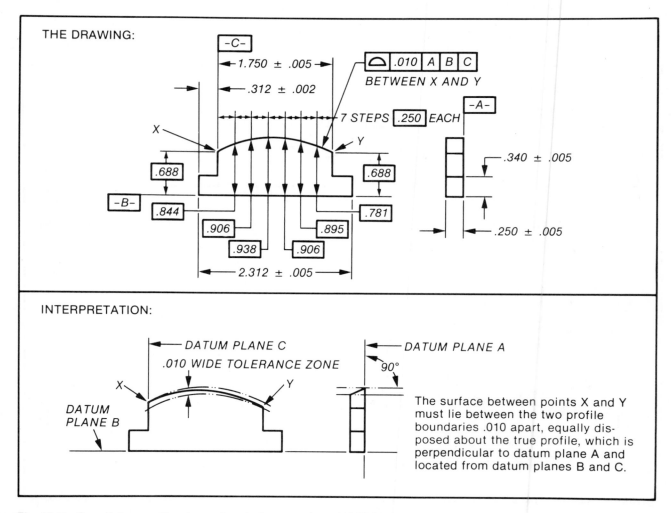

Fig. 11-2   Specifying profile of a surface between points. (ANSI Y14.5)

The tolerance may be applied on a bilateral or unilateral basis. Unless otherwise specified, the tolerance is assumed to be bilateral, half of it on the inside and half on the outside of the basic profile.

11.3 Specifying Profile Tolerances

An appropriate view is drawn showing the desired profile as a curved line. The desired profile is defined by basic dimensions, which may be located radii and angles, as in Fig. 11–3, or may be coordinate dimensions to points on the curve, as in Fig. 11–2.

To specify a bilateral tolerance, a feature control frame is drawn near the profile, with a leader pointing to the curved line [see Fig. 11–1 (a)]. When a unilateral tolerance is required, a portion of the boundary of the tolerance zone is drawn inside or outside the basic profile line, as applicable. The width of the tolerance is greatly exaggerated. It is usually drawn .06 inch (1.5 mm) wide to conform with the minimum distance between lines specified by DOD-STD-100. A dimension line indicating the width of the tolerance zone is then drawn and a feature control frame is attached to this dimension line, as shown in Fig. 11–1.

The extent of the profile to which the tolerance applies is specified on the drawing with identifying letters at limit points, as shown in Fig. 11–2. A note, such as "BETWEEN X & Y," is added below the feature control frame.

Different tolerances may be required on different portions of the profile. This is done by using separate feature control symbols with identifying letters at limit points. Fig. 11–3 is an example.

Where a profile tolerance applies all around the profile of a part, a circle is drawn on the bend of the leader, its diameter equal to the lettering height. This is the same symbol used to indicate "all around" in welding (see Fig. 11–4).

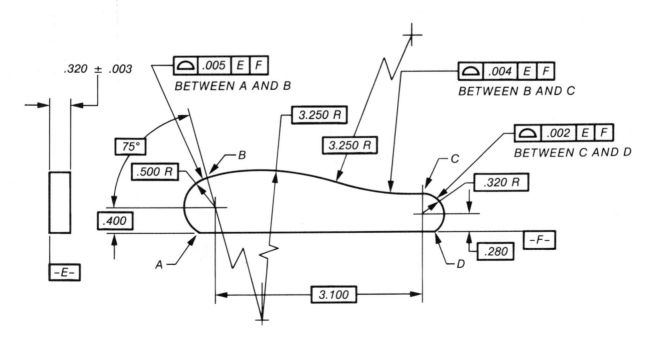

Fig. 11-3 Specifying different tolerances on different portions of the profile.

11.4   Application of Datums

Datums are usually necessary for profile of a surface tolerance in order to provide proper orientation of the profile. For profile of a line tolerance, datums may be used if the orientation of the profile is needed, but it is often not a requirement. For example, a datum is not specified when the only requirement is the proper shape at any particular cross section, as on a curved extrusion.

Note that Rule 1 (Section 6.2) does not apply to profile tolerances when used with a datum. Rule 1 applies only to individual features (not interrelated features) and states that no portion of a feature may extend beyond the size limits. When no datum is specified, Rule 1 does apply.

11.5   Profile Tolerance for Coplanar Surfaces

Two or more surfaces are coplanar if they are in the same plane. The two or more surfaces actually comprise one interrupted plane.

There is no special symbol for coplanarity. It can be controlled by a profile of a surface tolerance as shown in Fig. 11–5. The high points on the surfaces establish an implied datum from which other points on the surfaces may not vary by more than the specified tolerance. Notice that the leader from the feature control frame points to an extension line between the actual surfaces.

Every point on the coplanar surfaces must be within the size limits of the part (Rule 1, Section 6.2). The two surfaces are considered one interrupted plane and are not interrelated in their coplanarity to any other feature.

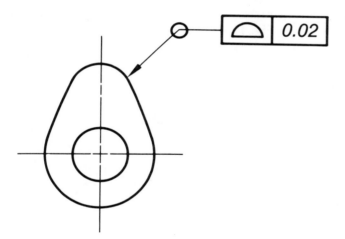

Fig. 11-4   An "all around" profile tolerance.

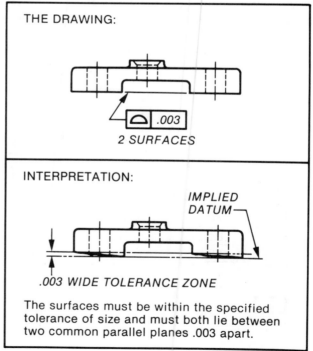

Fig. 11-5   Coplanarity controlled by a profile of a surface tolerance.

CHAPTER 12

PARALLELISM

12.1   Introduction

This study of geometric tolerance started with characteristics that do not require a datum. These characteristics are straightness, flatness, roundness, and cylindricity. They describe the required geometry all by themselves. They do not relate to any other feature. Next, profile tolerances were described. Profile tolerances can be used by themselves or can be related to another feature, a datum. All the remaining characteristics, starting with parallelism, relate the geometry of a feature to another feature and therefore require a datum.

12.2   Definition

Two surfaces are *parallel* if every point on each surface is the same distance from the other surface. *Parallelism error* is the measure of the amount two surfaces deviate from this condition.

Parallelism can also apply to circular features, such as cylinders and cones, in which case the *axes* are parallel. A cylinder, for example, may be parallel to another cylinder or to a flat surface.

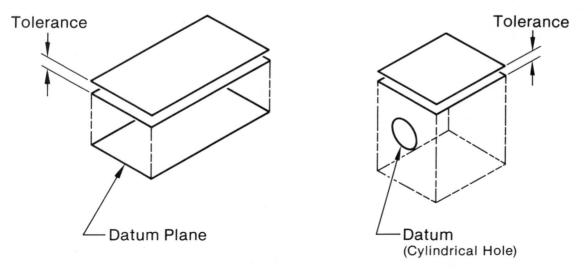

Fig. 12-1   Shape of tolerance zone when controlled feature is a flat surface.

12.3 Parallelism Tolerance Zone

The tolerance zone for parallelism may take either of two forms, depending on the form of the controlled feature and the datum.

(a) If the controlled feature is a flat surface, whether the datum is a flat surface or a circular feature, the tolerance zone will be the space between two imaginary planes, perfectly parallel to the datum, within which the flat surface must lie (see Fig. 12–1).

(b) If both the datum and the feature being controlled are circular features such as cylinders and cones, the tolerance zone will be an imaginary cylinder perfectly parallel with the axis of the datum. The axis of the controlled feature must lie within that cylinder. The tolerance zone is a cylinder because the axes of the two features must be parallel in all directions (see Fig. 12–2). In this case the tolerance is specified in the feature control frame as a diameter (see figs. 12–5 and 12–6).

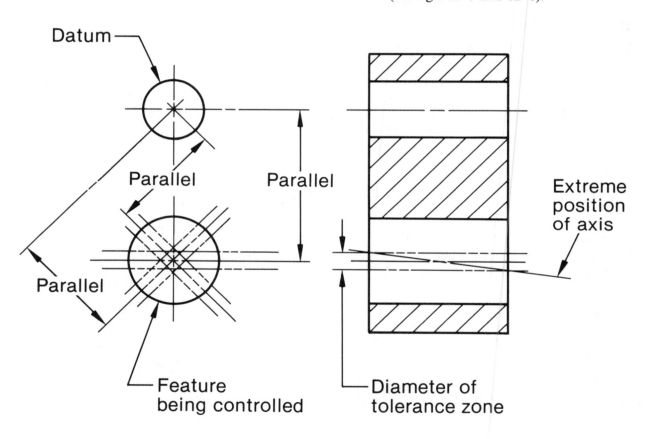

Fig. 12-2 Shape of tolerance zone when both controlled feature and datum are circular.

12.4 Specifying Parallelism Tolerance

Examples of how to specify parallelism tolerance are given in figs. 12–3 through 12–6.

Since parallelism is a relationship characteristic, a datum feature symbol is always required.

When features of size are involved, such as holes, the parallelism tolerance applies regardless of feature size (RFS). This is an application of Rule 3 (Section 6.4). When it is desired that the tolerance apply at MMC, this must be specified in the feature control frame. Fig. 12–6 is an example of how the tolerance is applied at MMC.

Rule 1 (Section 6.2) applies to all parallelism situations. The controlled feature must be within the tolerance limits of the location or size dimension.

12.5 Parallelism and Flatness

A parallelism tolerance for a flat surface also controls flatness when no flatness tolerance is specified. Every point on the actual surface must be within the parallelism tolerance zone; therefore, the surface must be flat at least within that tolerance.

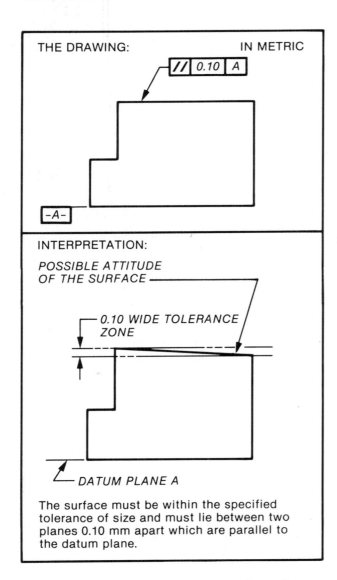

Fig. 12-3 Specifying parallelism of two flat surfaces.

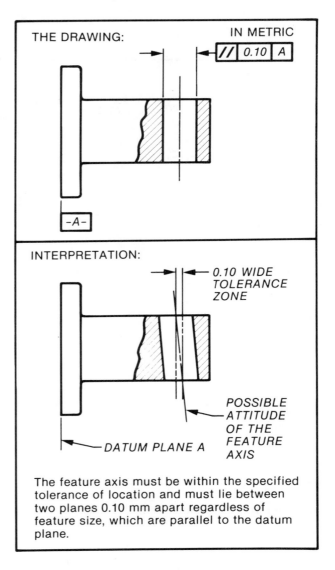

Fig. 12-4 Specifying parallelism of a circular feature and a flat surface.

THE DRAWING:

Ø.264–.267

// | Ø.005 | A

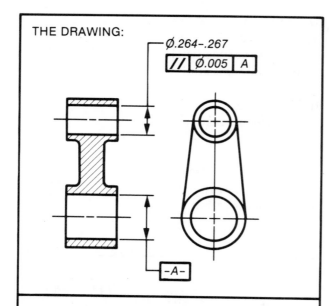

–A–

INTERPRETATION:

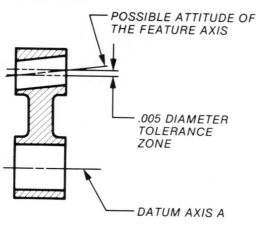

POSSIBLE ATTITUDE OF THE FEATURE AXIS

.005 DIAMETER TOLERANCE ZONE

DATUM AXIS A

The feature axis must be within the specified tolerance of location and must lie within a cylindrical zone .005 diameter, regardless of feature size, which is parallel to the datum axis.

Fig. 12-5 Specifying parallelism of two circular features—both features RFS. (ANSI Y14.5)

THE DRAWING:

Ø.264–.267

// | Ø.002 Ⓜ | A

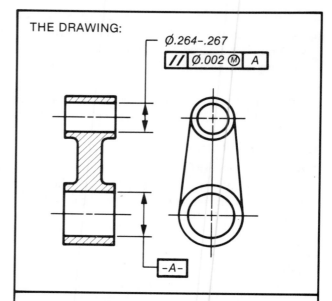

–A–

INTERPRETATION:

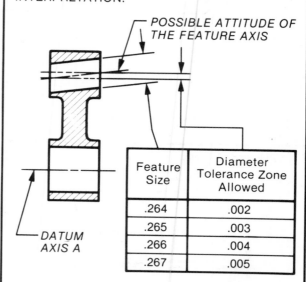

POSSIBLE ATTITUDE OF THE FEATURE AXIS

DATUM AXIS A

Feature Size	Diameter Tolerance Zone Allowed
.264	.002
.265	.003
.266	.004
.267	.005

The feature axis must be within the specified tolerance of location. Where the feature is at maximum material condition (.264), the maximum parallelism tolerance is .002 diameter. Where the feature departs from its MMC size, an increase in the parallelism tolerance is allowed which is equal to the amount of such departure.

Fig. 12-6 Specifying parallelism of two circular features—controlled feature MMC. (ANSI Y14.5)

CHAPTER 13

PERPENDICULARITY

13.1 Definition

Two plane surfaces or straight lines are *perpendicular* when they are at a right angle (90°) to each other. They do not necessarily have to intersect to be perpendicular. For example, the wall of your classroom is perpendicular to the floor of the next building. *Perpendicularity error* is any deviation in an actual part from a perfect right angle.

13.2 Perpendicularity Tolerance Zone

The tolerance zone may take any one of three forms, depending upon the form of the controlled feature and the datum.

(a) The most common form of the tolerance zone is the space between two imaginary parallel planes, perfectly perpendicular to the datum. This occurs under the following conditions:

  1. The controlled feature and the datum are both flat surfaces. The controlled surface must be within the parallel planes (see Fig. 13–1).

  2. The datum is a flat surface and the controlled feature is a slot or tab. In this case the median plane (an imaginary plane in the center) of the slot or tab must be between the parallel planes (see Fig. 13–2).

  3. The controlled feature and the datum are both circular features, in which case the axis of the controlled feature must be between the parallel planes (see Fig. 13–3).

(b) The tolerance zone may be in the form of an imaginary cylinder perfectly perpendicular to the datum. This occurs when the datum is a flat surface and the controlled feature is a circular feature, as shown in Fig. 13–4 and Fig. 13–5. The axis of the circular feature must be within the imaginary cylinder. The tolerance zone is a diameter rather than a width, as in other situations, because perpendicularity applies in all directions. An infinite number of parallel planes in all directions centered on an axis forms a cylinder. Note that the tolerance is specified as a diameter in the feature control frame.

(c) The least common form of the tolerance zone is the space between two imaginary parallel *lines,* perfectly perpendicular to the datum. Fig. 13–6 is an application of this, where a curved surface is perpendicular to a cylindrical size feature. Actually, the whole curved surface cannot be perpendicular, but each *radial element* is perpendicular to the datum. This is called radial perpendicularity. Below the feature control frame the note "EACH RADIAL ELEMENT" is added. Each radial element of the surface must be between the two parallel lines that make up the tolerance zone.

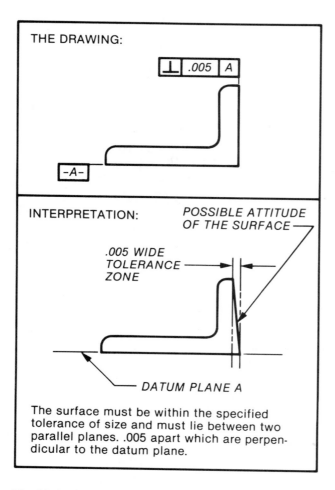

Fig. 13-1 Specifying perpendicularity of two flat surfaces.

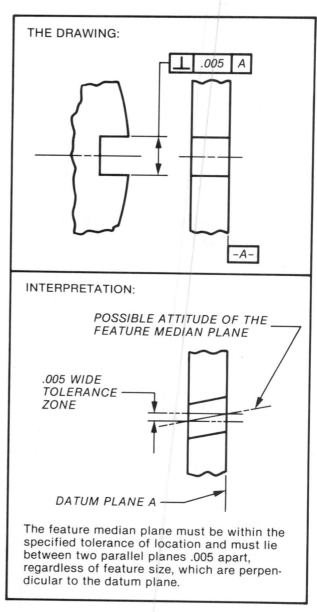

Fig. 13-2 Specifying perpendicularity for a median plane—the center of a slot. (ANSI Y14.5)

13.3 Specifying Perpendicularity Tolerance

Figs. 13–1 through 13–6 provide examples of how to specify perpendicularity on drawings. Since perpendicularity is a relationship characteristic, a datum is always required. Note that when the datum is a symmetrical size feature, the datum feature symbol is never placed on the center line. It is always given with the size dimension (Fig. 13–3).

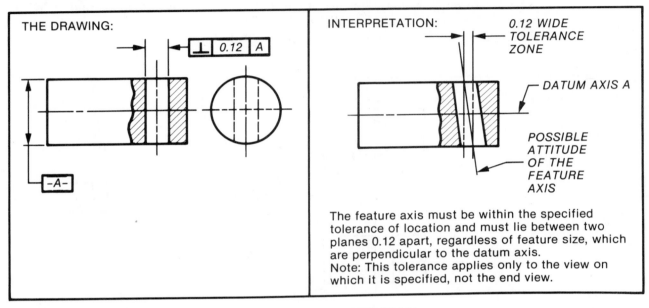

Fig. 13-3 Perpendicularity of two circular features (metric). (ANSI Y14.5)

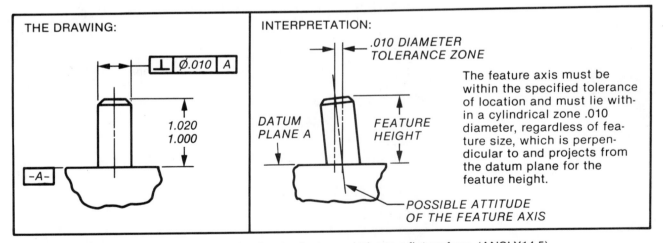

Fig. 13-4 Specifying perpendicularity of a circular feature relative to a flat surface. (ANSI Y14.5)

13.4 Use of Modifiers

Unless specified otherwise in the feature control frame, perpendicularity applies regardless of feature size (Rule 3, Section 6.4). Situations sometimes arise with size features where it might be desirable to have the tolerance apply at the maximum material condition. Fig. 13–7 is an example. Two mating parts are shown, one having a slot and the other a tab. In the worst tolerance condition, the tab must still fit into the slot.

In both of the parts shown in Fig. 13–7, the limits of the slot and the tab may exceed the dimensional limits. The tab, for example, may be as large as .370 (.365 maximum plus .005 perpendicularity tolerance). This seems to violate Rule 1, Section 6.2, which states that no part of a feature may extend beyond its MMC size. However, Rule 1 applies only to *individual* features. Since the tab and the datum surface are interrelated features, Rule 1 does not apply.

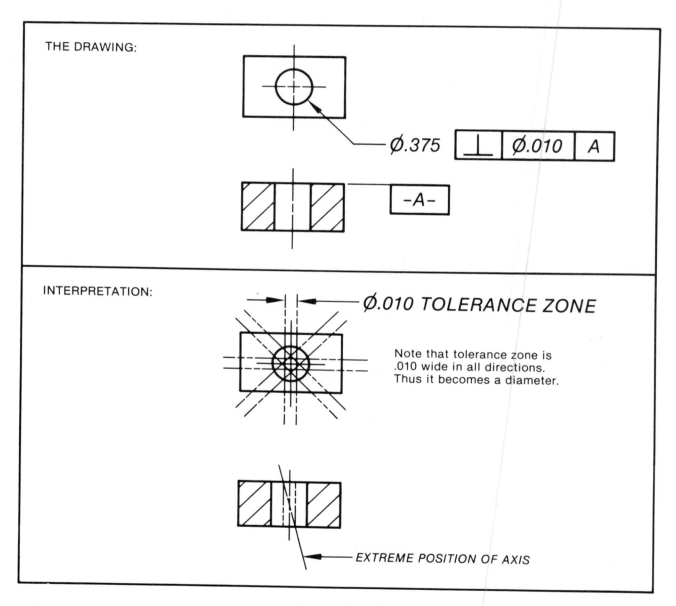

Fig. 13-5   For a cylinder perpendicular to a flat surface the tolerance zone is a diameter because the perpendicularity applies in all directions from the cylinder axis.

13.5   Perpendicularity and Flatness

A perpendicularity tolerance for a flat surface also controls flatness when no flatness tolerance is specified. Every point on the actual surface must be within the perpendicularity tolerance zone; therefore, the out-of-flatness cannot exceed the perpendicularity tolerance.

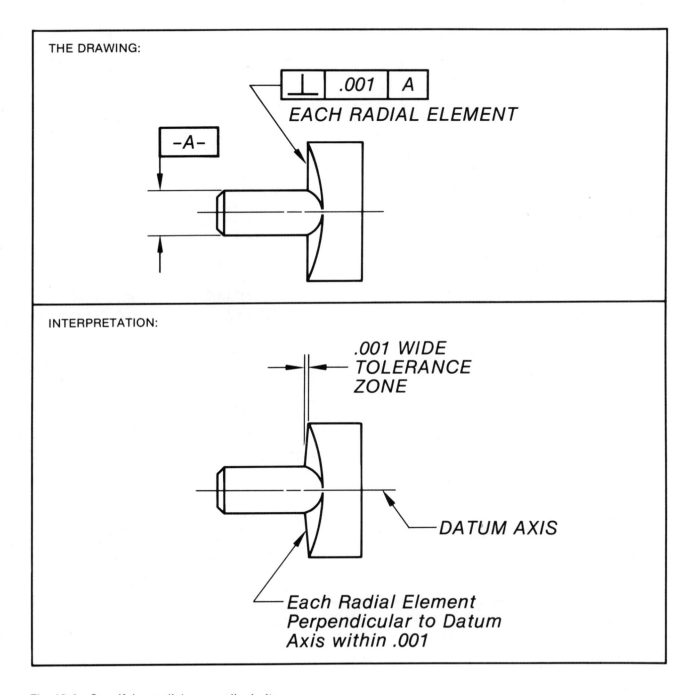

Fig. 13-6   Specifying radial perpendicularity.

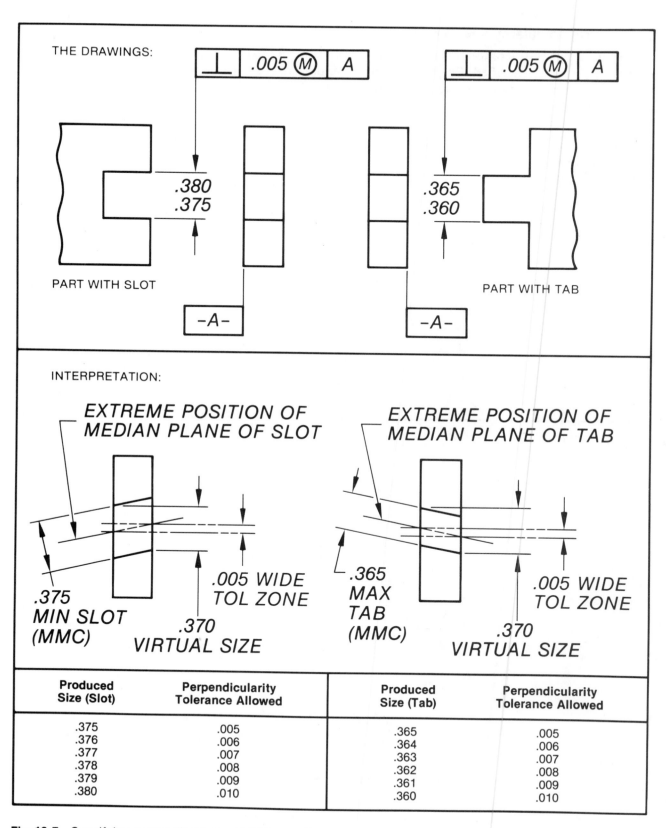

Fig. 13-7 Specifying perpendicularity at MMC— mating parts.

CHAPTER 14

ANGULARITY

14.1 Introduction

Angularity tolerance specifies the permissible error in an angle in a way different from the ordinary size tolerance attached to an angular dimension. A tolerance expressed in degrees results in a fan-shaped tolerance zone as shown in the example below.

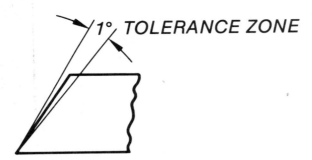

The tolerance is actually zero at the vertex and increases along the length of the angular surface. The increase for a 1° tolerance zone is about .017 inch per inch of length. For a 10-inch length (254 mm), it will be more than .170 inch (4.32 mm). This may not be acceptable in the design. It would be more desirable to have a tolerance zone of uniform width for the entire surface. This can be done by specifying the tolerance not in degrees but in thousandths of an inch or in parts of a millimeter, and this is exactly the function of geometric angularity tolerance. An angularity tolerance specifies the uniform width of a tolerance zone along the entire surface.

14.2 Angularity Tolerance Zone

The form of an angularity tolerance zone is always the space between two imaginary parallel lines or planes at an exact angle to a datum, within which every point of the controlled feature must lie. The feature may be a flat surface, or a circular feature, or a slot or tab. A datum must be specified from which the angle is to be measured. The datum, like the controlled feature, may be a flat surface, or the axis of a circular feature, or the median plane of a slot or tab.

Everything that has been said about angularity so far sounds much like perpendicularity, and it is. Actually, perpendicularity is a special case of angularity for an angle of 90°. Angularity is the general case for all other angles.

The angularity tolerance must be within the dimensional size limits for flat surfaces and within the dimensional location limits for size features (holes, slots, etc.). This is an application of Rule 1, Section 6.2.

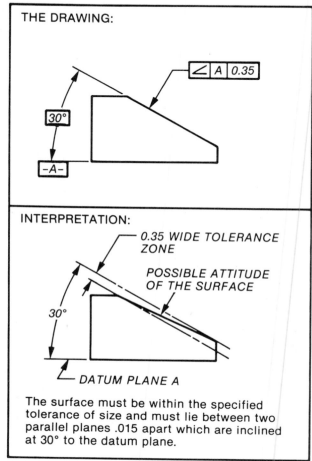

THE DRAWING:

30°

∠ A 0.35

-A-

INTERPRETATION:

0.35 WIDE TOLERANCE ZONE

POSSIBLE ATTITUDE OF THE SURFACE

30°

DATUM PLANE A

The surface must be within the specified tolerance of size and must lie between two parallel planes .015 apart which are inclined at 30° to the datum plane.

Fig. 14-1 Specifying angularity for a flat surface (metric). (ANSI Y14.5)

14.3 Specifying Angularity Tolerance

The angular dimension giving the required angle must be *basic* (without tolerance), since the tolerance will be specified in the feature control frame. The feature control frame is directed by means of a leader to the line representing the controlled feature or to an extension from it (see Fig. 14–1). Another method frequently used is to draw the feature control frame with a corner of it touching the extension line. This is shown in Fig. 14–2. When the feature being controlled is a size feature, the feature control frame may be drawn to the right of or below the size dimension, as shown in Fig. 14–3. Since angularity is a relationship characteristic, a datum is always required. Note that the feature control frame and datum feature symbol are never placed on the center line representing the axis or median plane of the feature.

14.4 Use of Modifiers

Unless otherwise specified in the feature control frame, angularity applies regardless of feature size (Rule 3, Section 6.4).

When dealing with size features, situations arise where it might be desirable to have the tolerance apply at the maximum material condition. A tab at some angle, for example, may have to fit into a slot in another part. However, this is best specified using a *positional* tolerance with a circle M modifier, which is explained in Section 18.8. Fig. 18–14 is an example of a positional tolerance used to control angularity.

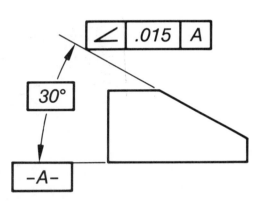

Fig. 14-2 Specifying angularity—alternate placement of the feature control frame.

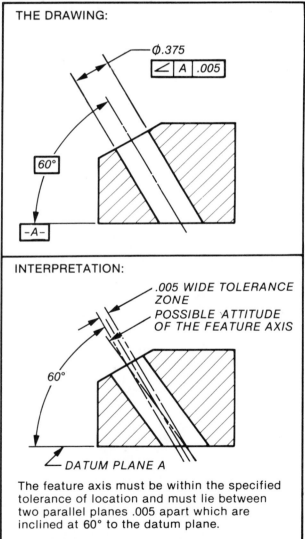

The feature axis must be within the specified tolerance of location and must lie between two parallel planes .005 apart which are inclined at 60° to the datum plane.

Fig. 14-3 Specifying angularity for a size feature. (ANSI Y14.5)

14.5 Angularity and Flatness

As with perpendicularity and parallelism, angularity also controls flatness of plane surfaces when no flatness tolerance is specified. Every point on the actual surface must be within the angularity tolerance zone; therefore, the out-of-flatness cannot exceed the angularity tolerance.

CHAPTER 15

CONCENTRICITY

15.1 Coaxiality

Two cylinders are *coaxial* if they have the same axis or if their axes are on one line.

Strictly speaking, circles are not coaxial, since they have no axes; they have centers. Circles on the same center are concentric. Cylinders and other regular figures do have axes and should be termed coaxial when they have a common axis. However, in engineering practice, the terms *coaxial* and *concentric* are used interchangeably.

Any two *regular* shapes that are symmetrical about an axis, such as cylinders, cones, curved profiles, squares, hexagons, and so forth, may be coaxial (see Fig. 15-1). This discussion deals mostly with cylinders, but all of the information applies equally well to other regular shapes.

The word *coaxiality* is used to describe the general case of the geometric characteristic where axes are in line. There is no geometric characteristic symbol for coaxiality. Instead, any of three symbols that define special cases are used. These are concentricity, runout, and positional tolerance. The difference is principally in the form of the tolerance zone. This chapter will examine concentricity. Chapter 16 will deal with runout and chapters 17, 18, and 19 with positional tolerance.

15.2 Concentricity Tolerance

Concentricity error is the amount by which the axes of two regular solids are out of line. However, it is not measured as the distance between the axes. Error in concentricity is sometimes referred to as *eccentricity* (literally "off-centeredness"), but this term is generally used in connection with circular shapes that are off-center by design.

Concentricity is more restricted in its use than the other coaxiality characteristics and it is therefore specified least often.

15.3 Concentricity Tolerance Zone

The form of the tolerance zone is an imaginary cylinder about the exact axis of the datum. The diameter of this cylinder is equal to the specified tolerance. The axis of the controlled feature must lie within the imaginary cylinder. This is shown in figs. 15-2 and 15-3. Notice that the axis of the controlled feature may be

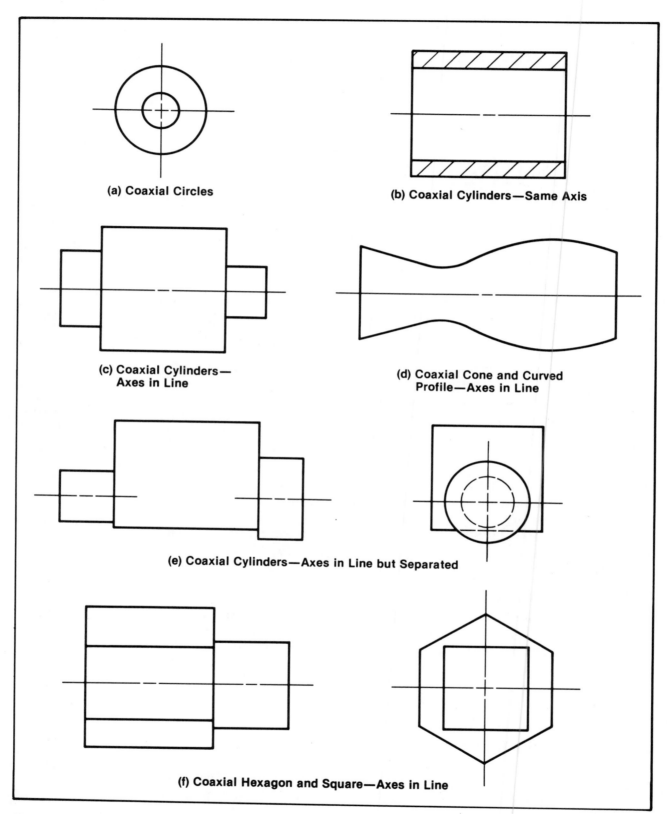

(a) Coaxial Circles

(b) Coaxial Cylinders—Same Axis

(c) Coaxial Cylinders—
Axes in Line

(d) Coaxial Cone and Curved
Profile—Axes in Line

(e) Coaxial Cylinders—Axes in Line but Separated

(f) Coaxial Hexagon and Square—Axes in Line

Fig. 15-1   Examples of coaxiality.

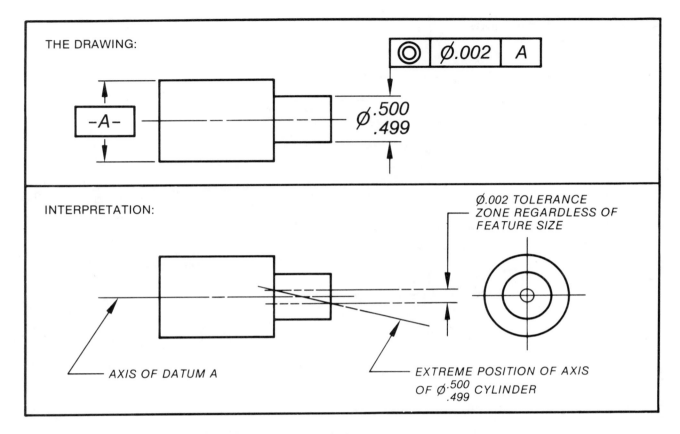

THE DRAWING:

INTERPRETATION:

Ø.002 TOLERANCE ZONE REGARDLESS OF FEATURE SIZE

AXIS OF DATUM A

EXTREME POSITION OF AXIS OF Ø$^{.500}_{.499}$ CYLINDER

Fig. 15-2   Specifying concentricity—datum a single cylinder.

anywhere within the tolerance zone. Therefore, it can be slanted relative to the datum axis. This is one reason why concentricity error is not measured as the distance between the datum axis and the feature axis. Actually, if the feature axis is slanted, this is not concentricity error but parallelism error. However, the slant is treated as an error in concentricity.

15.4   Specifying Concentricity

Concentricity is specified as shown in Fig. 15–2 and Fig. 15–3. Since the form of the concentricity tolerance zone is a cylinder, the tolerance is specified as a diameter. This is done by placing the diameter symbol (Ø) *before* the tolerance.

A datum is always necessary since concentricity is a relationship characteristic like parallelism and perpendicularity. It is important to select as datums surfaces that are functional in the use of the part. In a cylindrical part this might be a diameter that fits into a bearing or mating part, or it might be a flat face that locates the part in an assembly. Lathe centers are never used since they have no function in the use of the part.

The datum may be a single cylinder, as in Fig. 15–2, or it may be two separated cylinders used as one datum. This is shown in Fig. 15–3. In this situation the object is rested on both cylinders for concentricity testing of the controlled diameter. The two cylinders comprise one datum. This is indicated in the feature control frame by connecting the two datum letters with a dash rather than separating them with a vertical line.

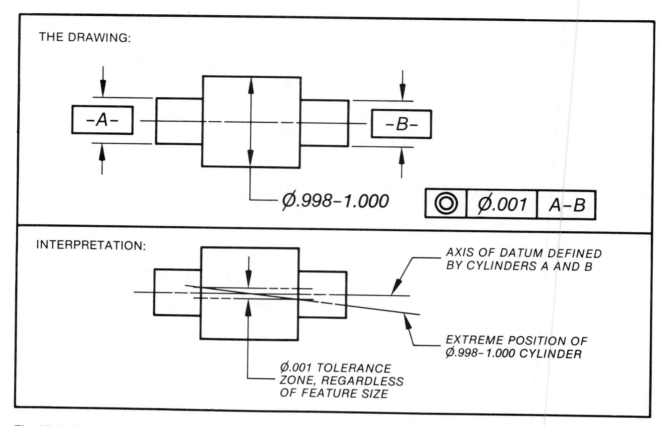

Fig. 15-3 Specifying concentricity—datum defined by two separated cylinders.

Concentricity does not include errors in roundness and straightness. Because it is difficult to separate these errors from true concentricity error, concentricity is only specified when this characteristic is essential in the design.

Generally, concentricity is specified when coaxiality must be controlled independently of surface errors of roundness and straightness. An example is a motor rotor in which excessive coaxiality error would have a large effect on balance. The part shown in Fig. 15–3 might be such a rotor. Surface errors, which have little effect on balance, can be controlled with separate feature control symbols or perhaps only by tight size tolerances.

15.5 Use of Modifiers

Concentricity tolerance always applies regardless of the size of the controlled feature and the datum (RFS). If a fit at the maximum material condition (MMC) is required, positional tolerance is preferred.

15.6 Testing for Concentricity Error

Since it is not possible to get inside an object to measure the distance between axes of cylinders (especially since the axes are only imaginary), concentricity must be tested from the outside. If both the datum and the controlled feature were perfectly round and straight, the datum could be rotated in a vee-block while holding the probe of a dial indicator against the controlled cylinder. The full indicator movement (FIM) would be exactly equal to the diameter of the concentricity error. However, perfect roundness and straightness are never achieved, so part of the FIM in actual practice is due to these errors. To eliminate unwanted errors in the reading, independent tests have to be made for roundness and straightness of both the datum and the controlled feature; then these are subtracted from the original reading.

If the FIM in a vee-block/dial indicator test is within the concentricity tolerance specified on the drawing, the part is acceptable even though part of the reading is due to surface errors.

When it is necessary to obtain pure concentricity readings without recording surface errors, tests can be made using special fixtures. For information on these methods, the student is referred to texts on precision inspection methods.

15.7   Selection of Proper Control for Coaxiality

Concentricity: Use when

1.   Coaxiality must be controlled independently of surface errors.

     Separate feature control symbols may be used to control surface errors.

2.   The desired coaxiality tolerance must be held regardless of feature and datum size (RFS).

Runout: Use when

1.   Surface errors may be included with coaxiality error.

     No separate inspection will be made for surface errors unless specified on the drawing.

2.   The desired coaxiality tolerance must be held regardless of feature and datum size (RFS).

Positional Tolerance: Use when

1.   Surface errors may be included with coaxiality error.

     No separate inspection will be made for surface errors unless specified on the drawing.

2.   The coaxial features must assemble with another part having corresponding coaxial features, and additional coaxial tolerance may be allowed when the feature and/or the datum are not at the maximum material condition (see Section 19.4).

CHAPTER 16

RUNOUT

16.1 Introduction

Runout is any deviation of a surface from perfect form that can be detected by rotating the part about an axis. Runout tolerance is the maximum deviation allowed. It is a composite tolerance including errors in roundness, straightness, perpendicularity, and coaxiality. All of these geometric errors can be read as runout if they are detected by rotating the part about an axis.

Runout may be applied to any surface generated around an axis, such as cylinders, cones, and curved profiles. It is also applied to flat faces perpendicular to an axis.

16.2 Circular and Total Runout

There are two types of runout control: circular runout and total runout.

In circular runout the deviation of each circular element is controlled; there is no control over elements in any other direction.

Total runout controls the deviation of all elements of a surface, circular or straight. The entire surface is controlled.

Circular runout is less expensive to measure and is adequate for most design functions. It is therefore more commonly specified than total runout.

16.3 Runout Tolerance Zone

The shape of the tolerance zone is different for circular and total runout, and for each type of control it can take any of four forms, depending upon the shape of the feature being controlled. All of these forms are tabulated and illustrated in figs. 16-1 and 16-2. Notice that for circular runout the tolerance is the distance between theoretical *elements,* while for total runout it is the distance between theoretical *surfaces.*

16.4 Specifying Runout

Runout tolerance is specified in the same way as other geometric tolerances expressing a relationship of one feature to another. One or more datums are identified on the drawing and specified in the feature control frame. Examples of both types of runout control are shown in figs. 16-3 through 16-8. The double arrow symbol for total runout is derived from the fact that the surface must be tested in two directions (circular elements and straight elements).

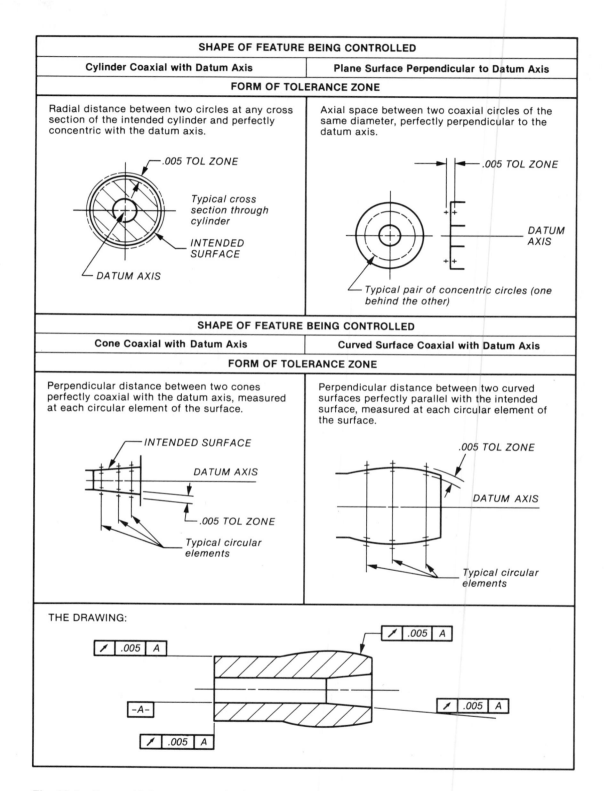

Fig. 16-1 Form of tolerance zone for four types of features—circular runout.

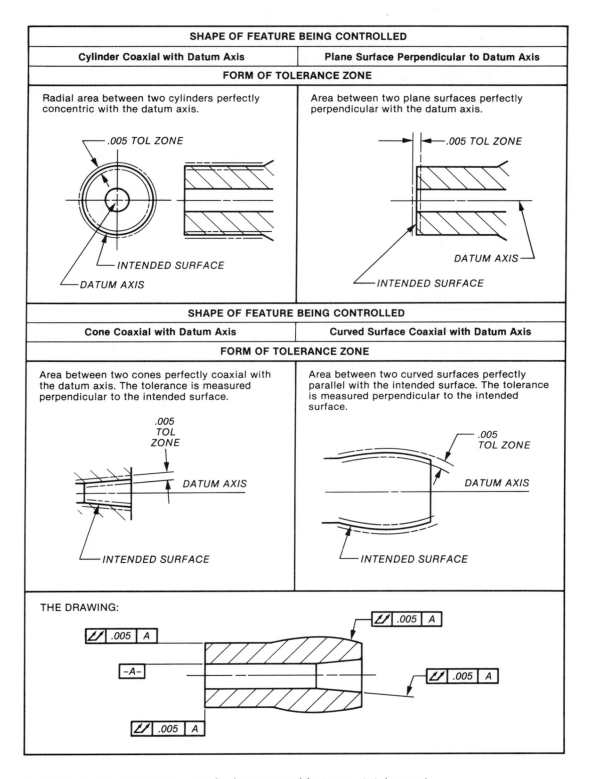

Fig. 16-2   Form of tolerance zone for four types of features—total runout.

As with most geometric characteristics, runout may be combined with other tolerances. For example, a diameter with a .05 mm runout tolerance may have to be held cylindrical within .02 mm. This is done by attaching a cylindricity feature control frame to the runout frame (see Fig. 16–8).

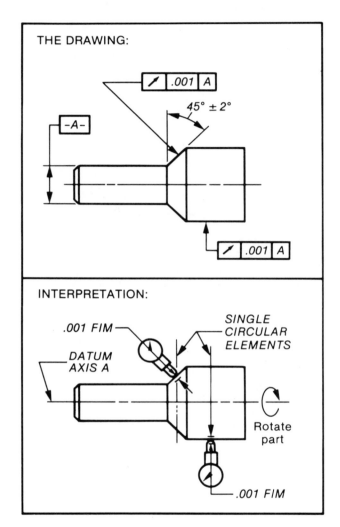

Fig. 16-3 Circular runout—datum a single cylinder.

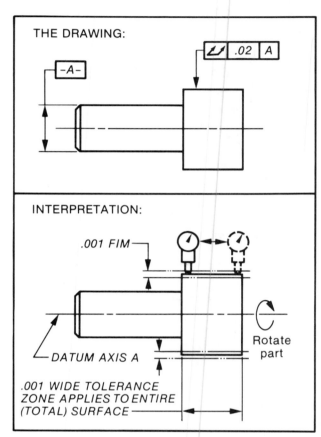

Fig. 16-4 Total runout—datum a single cylinder.

16.5 Application of Datums

As in concentricity tolerance, it is important to select as datums surfaces that are important in the function of the part. In a cylindrical part this might be a diameter that fits into a bearing or other mating part, or it might be a flat face that locates the part in an assembly. Lathe centers are never used since they have no function in the use of the part.

The datum may be a single cylinder as in figs. 16–3 and 16–4, or it may be two separated cylinders used as one datum. This is shown in Fig. 16–5. In this situation the object is rested on both cylinders for testing runout of the controlled

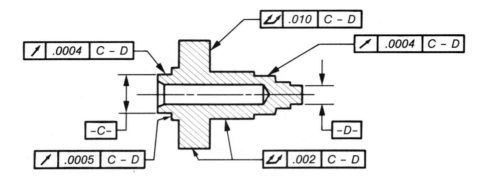

Fig. 16-5 Circular and total runout—datum defined by two separated cylinders.

feature. The two cylinders comprise one datum. This is indicated in the feature control frame by connecting the two datum letters with a dash rather than separating them with a vertical line.

Where cylindrical parts are relatively large in diameter and short in length, it is necessary to specify a flat face as a datum in addition to a cylindrical surface. This is illustrated by the parts in figs. 16–6 and 16–7. In this case the part would tilt if laid in a vee-block for testing, so it is steadied by holding the flat face against appropriate tooling perpendicular to the vee.

16.6   Use of Modifiers

Runout tolerance always applies regardless of the size of the feature and the datum (RFS). If a fit at the maximum material condition (MMC) is required, this may be accomplished by specifying positional tolerance instead of runout.

16.7   Testing for Runout Error

Runout error is commonly measured by contacting the surface with a dial indicator while the part is rotated 360° in a vee-block (for an external cylinder datum) or over a mandrel (for an internal cylinder datum). The full indicator movement (FIM) is equal to the runout error.

When a flat face is being inspected, it is necessary to place a stop against the opposite face of the part to prevent axial movement that would also register on the indicator (see Fig. 16–7).

In testing for circular runout on both cylindrical and plane surfaces only circular elements are inspected. The body of the dial indicator is not allowed to move during each rotation of the part. Enough circular elements are inspected to ensure that the readings are typical of all circular elements (see Fig. 16–3).

When total runout is being tested, several circular elements are inspected; then several longitudinal elements (on cylinders) or radial elements (on flat faces) are inspected. The latter inspections are done by moving the dial indicator longitudinally or radially as required while the part is held still. After each inspection the part is rotated to a new position and another element is tested (see Fig. 16–4).

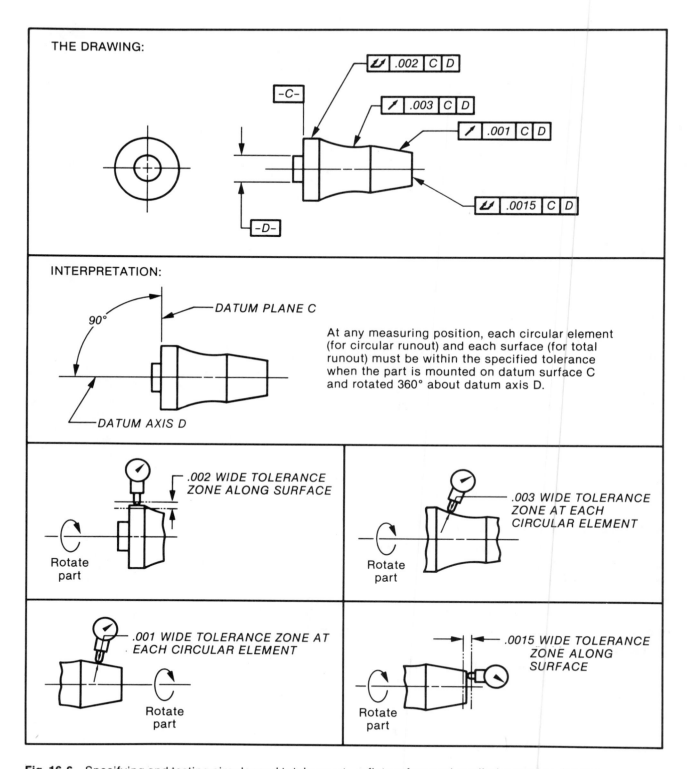

THE DRAWING:

INTERPRETATION:

DATUM PLANE C

90°

DATUM AXIS D

At any measuring position, each circular element (for circular runout) and each surface (for total runout) must be within the specified tolerance when the part is mounted on datum surface C and rotated 360° about datum axis D.

.002 WIDE TOLERANCE ZONE ALONG SURFACE

Rotate part

.003 WIDE TOLERANCE ZONE AT EACH CIRCULAR ELEMENT

Rotate part

.001 WIDE TOLERANCE ZONE AT EACH CIRCULAR ELEMENT

Rotate part

.0015 WIDE TOLERANCE ZONE ALONG SURFACE

Rotate part

Fig. 16-6 Specifying and testing circular and total runout—a flat surface and a cylinder used as datums.

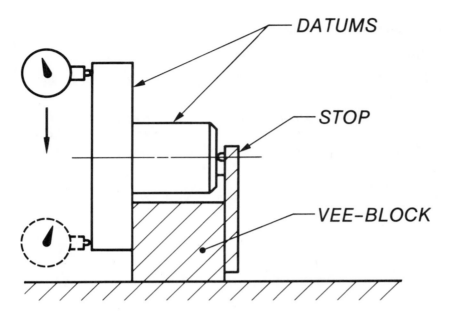

Fig. 16-7   A flat-face datum held against tooling for testing runout.

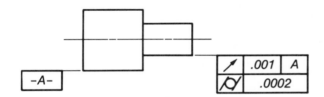

Fig. 16-8   Combined runout and cylindricity tolerance.

16.8   Selection of Proper Control for Coaxiality

Runout: Use when

1.   Surface errors may be included with coaxiality error.

   No separate inspection will be made for surface errors unless specified on the drawing.

2.   The desired coaxiality tolerance must be held regardless of feature and datum size (RFS).

Concentricity: Use when

1.   Coaxiality must be controlled independently of surface errors.

   Separate feature control symbols may be used to control surface errors.

2.   The desired coaxiality tolerance must be held regardless of feature and datum size (RFS).

Positional Tolerance: Use when

1. Surface errors may be included with coaxiality error.

   No separate inspection will be made for surface errors unless specified on the drawing.

2. The coaxial features must assemble with another part having corresponding coaxial features, and additional coaxiality error may be allowed when the feature and/or the datum are not at the maximum material condition (see Section 19.4).

CHAPTER 17

POSITIONAL TOLERANCE— GENERAL

17.1 Definition

Until 1973, when the name was changed to conform with international usage, *positional tolerance* was known as true position tolerance. The older term will be useful in describing what positional tolerance is.

The true position of a feature is the theoretically exact location of its axis or center plane from the feature or features from which it is dimensioned.

Any deviation from the true position was formerly called true position error and is now known as *positional error*. Positional tolerance, therefore, is the total permissible error in the location of a feature relative to another feature or to several other features.

17.2 Positional Tolerance

Positional tolerance is a composite tolerance. It is composed of several different errors, which could include errors in roundness, straightness, parallelism, and perpendicularity of a feature in addition to mislocation of its axis or center plane from the true position.

Positional tolerance can be used to control the location of the following types of features.

1. Holes, circular and noncircular.
2. Bosses or other protruding features.
3. Slots, notches, and tabs.
4. Cylinders, cones, and other regular shapes on a common axis (coaxiality).

Positional tolerance is the most versatile and the most commonly used of the geometric characteristic tolerances.

17.3 Positional Tolerance Zone

The form of the tolerance zone for positional tolerance depends upon whether the feature is located on an axis or a center plane.

For a feature located by its axis, such as a round hole, the tolerance zone is an imaginary cylinder perfectly parallel, perpendicular, or concentric to the datum, as required, and centered on the true position of the axis (see Fig. 17–1). The diameter of the cylinder is equal to the specified tolerance; its length is equal to the length of the feature. The axis of the feature as produced must lie entirely within the cylinder. The tolerance value given in the feature control frame is always specified as a diameter. For example, in Fig. 17–6, it is ⌀.014.

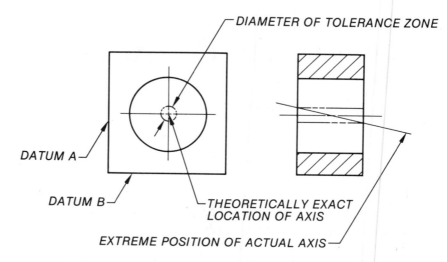

Fig. 17-1 Positional tolerance zone for a feature located by its axis.

For a feature located by its center plane, such as a rectangular tab, the tolerance zone is the space between two imaginary planes perfectly parallel to and equidistant from the true position of the center plane (see Fig. 17–2). The distance between the imaginary planes is equal to the specified tolerance. The length of the tolerance zone is equal to the length of the feature. The center plane of the feature as produced must lie between the imaginary planes. The tolerance specified in the feature control frame is the *total* width of the tolerance and not the distance from the center plane of the datum to either side of the tolerance zone.

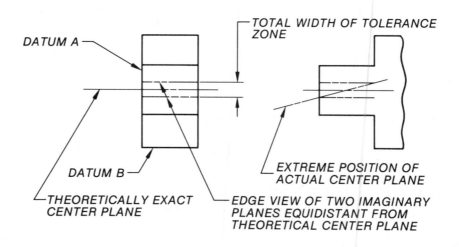

Fig. 17-2 Positional tolerance zone for a feature located by its center plane.

17.4 Comparison of Coordinate Tolerancing with Positional Tolerancing

A familiar problem in mechanical design is the dimensioning of features such as holes that are perpendicular to a surface. This is commonly done by *coordinate dimensions,* dimensions in two perpendicular directions. The dimensions may be given from hole to hole (chain dimensioning) or all from a single surface in each of the two directions (base-line or datum dimensioning). Every dimension is assigned a tolerance, usually bilateral (±), or it may be in the form of a limit dimension. Because of the accumulation of tolerances where there are more than two holes, chain dimensioning is not used for precision manufacturing. Therefore, only base-line coordinate dimensioning will be considered. An example of this is given in Fig. 17–3.

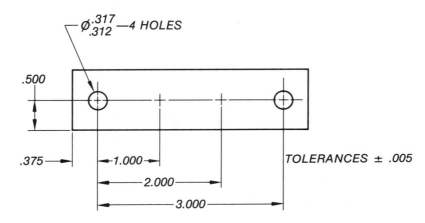

Fig. 17-3 Base-line coordinate dimensioning.

For the part shown in Fig. 17–3, the tolerance zone for the location of each hole is a square, .010 × .010, since all the tolerances are ±.005. One such tolerance zone is shown in Fig. 17–4. If the tolerance of one of the dimensions were different from ±.005, the tolerance zone would be rectangular. For example, if the vertical tolerance were ±.008, the tolerance zone would be .016 high by .010 wide.

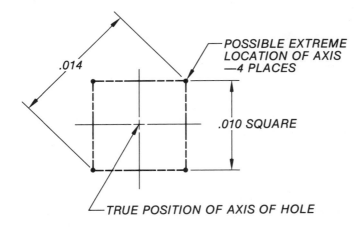

Fig. 17-4 Enlarged view of tolerance zone for one hole.

Looking again at Fig. 17–4, notice that if the actual axis of the hole were to be in any of the four corners, it would be .007 from the true position, and the total actual error could be as large as .014. Now suppose the actual axis was found to be .007 to the left or right of the true position, or .007 above or below. This is no worse than the .007 diagonal error, yet the part would be rejected because it is not "in tolerance."

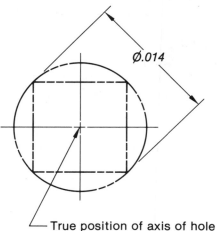

Fig. 17-5   Equivalent positional tolerance zone.

Using positional tolerance, the tolerance zone is a circle centered on the true position of the axis, as shown in Fig. 17–5. In this example the diameter of the tolerance zone is the same as the diagonal measurement of the coordinate tolerance zone, yet its area is much larger—actually 57 percent larger. (The proof of this is left to the student.) This means that if positional tolerance is specified there is less likelihood of rejected parts, with the same assurance that the parts will fit together. And this is the beauty of positional tolerancing.

17.5 Specifying Positional Tolerance

Fig. 17–6 shows the same part as in Fig. 17–3 with the holes controlled by a positional tolerance. All of the location dimensions are *basic* dimensions (without

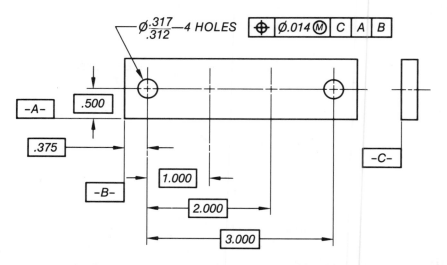

Fig. 17-6   Holes located by basic dimensions and controlled by positional tolerance.

tolerance), since they now give only the desired true position of the holes. The tolerance is given in the feature control frame, which states that the holes are to be in true position relative to the three datums within a .014-diameter tolerance zone.

Three possible effects of positional tolerance on the actual axis of a feature are shown in Fig. 17–7. Notice that the positional tolerance also controls the perpendicularity of the holes relative to the surface of the part. The axis produced can be slanted by .014 within the thickness of the part.

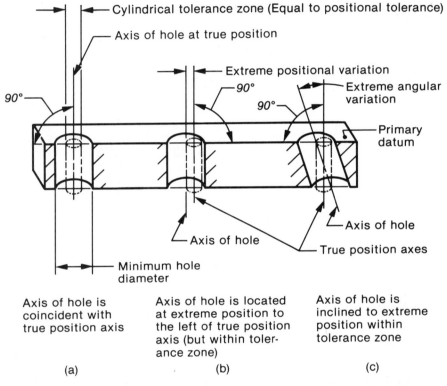

Note that the length of the tolerance zone is equal to the length of the feature, unless otherwise specified on the drawing.

Fig. 17-7   Three possible effects of positional tolerance on the axes of holes.

17.6 Datums for Positional Tolerance

Some of the facts learned about datums in Chapter 4 will be reviewed here and applied to positional tolerance.

In Chapter 4 advice on which features to select as datums was given in the form of three rules. They will be repeated here.

1. Select datums that are functional features. A cylindrical surface that supports a part in a bearing is a functional surface and should be used to control other features. Lathe centers are not used as datums because they do not function in the use of the part.

2. Select corresponding features on mating parts, features that fit together.

3. Select features that are readily accessible for manufacturing and inspection.

In positional tolerancing one or more datums must be specified for every feature being controlled. An example of a part with a hole pattern related to three datums is shown in Fig. 4–3. The central hole is selected as the first datum (A) because it fits over the mating part and is essential in locating the hole pattern. The second and third datums (B and C) provide the correct angular orientation and control the squareness of the hole pattern. Other examples of the use of three datums are figs. 17–6 and 17–8.

In many cases the datum itself must have its geometric form controlled. For example, if the datum is a plane surface, flatness may have to be specified; if it is a cylinder, roundness or cylindricity may be necessary. This is done with separate feature control symbols applied to the datum. Fig. 17–8 is an example.

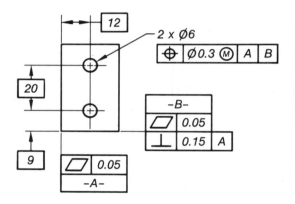

Fig. 17-8   Datum B has its own geometric form control (metric).

When a datum is a size feature (Section 3.1), complications arise because of the dimensional tolerance of the size. This size variation must be considered and a decision made whether the datum is to apply at MMC, LMC, or RFS. For positional tolerance the modifier must always be specified (Rule 2, Section 6.3).

For most applications, it will be desirable to let the datum apply at MMC so that additional tolerance can be permitted when the produced size is not the maximum material size (see Fig. 6–3). An analysis of the advantages of MMC is given in Section 17.7.

There is an additional complication when a size feature is used as a datum and it has its own geometric tolerance. The datum feature size applies at its *virtual* condition (the size that produces the tightest fit with a mating part) even though MMC is specified. This is an application of Rule 5 (Section 6.6). Fig. 6–3 is an example.

Positional tolerance is not often specified with the LMC modifier. A typical application is explained in Section 17.7 under "Least Material Condition."

When the positional tolerance must be held as specified no matter how big or small the datum, the modifier Ⓢ is placed after the datum reference letter in the feature control frame as shown in Fig. 17–9.

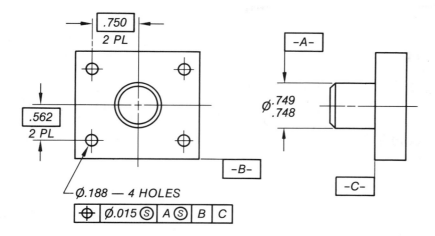

Fig. 17-9   Datum A, a size feature, has its own modifier.

17.7   Use of Modifiers

Any of the three modifiers may be specified with positional tolerances: MMC, LMC, and RFS.

Maximum Material Condition (MMC) Ⓜ

It will be recalled from Rule 2 (Section 6.3) that a modifier must always be specified for a positional tolerance. In Fig. 17–6, MMC is specified. This means that the specified tolerance applies only when the feature is at MMC, and as the produced size departs from MMC, more tolerance can be added.

Now suppose the size of one of the holes as drilled or punched were .317 (.005 over MMC). This .005 can be added to the positional tolerance of .014, giving a more liberal tolerance of .019. A table could be made up comparing various produced sizes with resulting positional tolerances. The following is an example.

Produced Size	Positional Tolerance
.312	.014
.313	.015
.314	.016
.317	.019

This extra tolerance is often called a *bonus tolerance* and it is aptly named, since it is received free, gratis, no extra charge.

Least Material Condition (LMC) Ⓛ

Situations arise when it is desirable to stipulate that a positional tolerance apply at the least material condition (LMC), the condition where a solid feature is the minimum size and a hole is the maximum size—where in each case the part contains the least material. An example is given in Fig. 17–10. The symbol Ⓛ

("circle L") is used both for the hole location and the datum, the outside diameter (O.D.). In this object the edge distance (E.D.) between the edge of the hole and the O.D. is critical. The minimum E.D. resulting from the two size tolerances is 3.93 mm, which can be decreased to 3.68 mm by the positional tolerance of ⌀0.5 (R 0.25). When the hole is produced smaller and the O.D. larger, departing from their LMC, more positional error can be allowed without affecting the critical E.D.

As with Fig. 17–6, a table can be made up comparing various produced sizes with resulting positional tolerances.

Produced Size (mm)		Total Departure from LMC (mm)	Permissible Positional Error (mm)	Min. Edge Distance (mm)
Holes	Datum			
25.03	76.9	0	⌀0.5	3.68
25.02	77.0	0.11	⌀0.61	3.68
25.01	77.1	0.22	⌀0.72	3.68
25.00	77.1	0.23	⌀0.73	3.68

Following is a sample calculation for the produced sizes of 25.00 mm for the holes and 77.1 mm for the outside diameter.

$$\begin{array}{rl}
& \varnothing25.03 \text{ LMC size of hole} \\
- & \underline{\varnothing25.00} \text{ actual size of hole} \\
& .03 \text{ departure from LMC}
\end{array}$$

$$\begin{array}{rl}
& \varnothing77.1 \text{ actual size of outside diameter} \\
- & \underline{\varnothing76.9} \text{ LMC size of outside diameter} \\
& .2 \text{ departure from LMC} \\
+ & \underline{\quad.03} \\
& .23 \text{ total departure from LMC} \\
\varnothing & \underline{\quad.5} \text{ positional tolerance at LMC} \\
\varnothing & .73 \text{ total permissible positional tolerance}
\end{array}$$

$$\begin{array}{rl}
& 44.00 \text{ location of holes, axis to axis (basic)} \\
+ \varnothing & \underline{\quad.73} \text{ total permissible positional tolerance} \\
& 44.73 \\
+ & \underline{\varnothing25.00} \text{ actual hole size} \\
& 69.73 \text{ edge of hole to far edge of opposite hole}
\end{array}$$

$$\begin{array}{rl}
& \varnothing77.10 \text{ actual outside diameter} \\
- & \underline{\quad69.73} \\
& 7.37 \text{ edge distance, both holes}
\end{array}$$

$7.37 \div 2 = 3.68$ edge distance each hole

Again there is a *bonus tolerance*, this time totaling 0.23 mm when both the hole and the O.D. are farthest from LMC.

Regardless of Feature Size (RFS) Ⓢ

Where there is a reason to hold the location of a feature regardless of its size (RFS), this can be done by adding the symbol Ⓢ ("circle S") after the tolerance in the feature control symbol. An example is shown in Fig. 17–9. Note that in this object one of the datums is a *size* feature and Ⓢ is applied to the datum as well as to the holes. This means that no matter how big or small the datum and the holes, the positional tolerance is still .015 diameter.

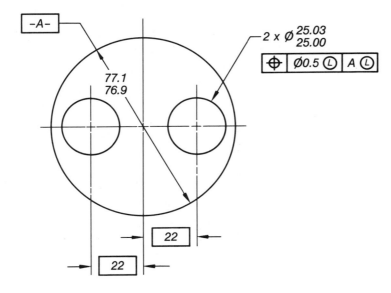

Fig. 17-10   An application of LMC to maintain a critical edge distance (metric).

CHAPTER 18

POSITIONAL TOLERANCE—LOCATION APPLICATIONS

18.1 Mating Parts Assembled with Fasteners

The advantage of positional tolerance shows up best when applied to mating parts fastened with bolts, screws, pins, and studs. Such parts may be held together in two different ways.

a. Floating Fasteners (Fig. 18–1)
Both parts have plain cylindrical holes (clearance holes) that are larger than the fasteners. Since the fasteners are free to float within the clearance holes, an assembly of this kind is called *floating fasteners*.

b. Fixed Fasteners (Fig. 18–4)
One part has a clearance hole and the other has a tight-fitting hole that holds the fastener in a fixed position. The tight hole may be a press fit for a pin or it may be a tapped hole to accommodate a screw or stud. This type of assembly is known as *fixed fasteners*.

In the design of mating parts assembled with fasteners it is necessary to make decisions about the amount of clearance required in the holes and the positional tolerance to be specified. The two are related. For floating fasteners the tolerance is equal to the clearance and for fixed fasteners the tolerance is equal to half the clearance.

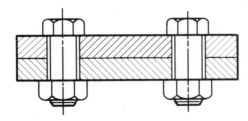

Fig. 18-1 Floating fastener assembly.

The formulas given below will produce a metal-to-metal fit when the holes and fasteners are at MMC and the hole axes are at the extremes permitted by the positional tolerance. When holes and fasteners as produced are not at their MMC, there will be more clearance—or additional positional error can be allowed.

The following symbols are used.

F = Max Dia of *Fastener* (MMC)

H = Min Dia of *Hole* (MMC)

T = Dia of Positional *Tol*

18.2 Analysis of Floating Fasteners

Once a decision has been made on the fastener size required, the hole size and positional tolerance can be calculated by the following formulas.

$$H = F + T \qquad T = H - F$$

From the second formula it can be seen that the tolerance (T) is equal to the clearance (H – F).

Example: Assume that the fastener in Fig. 18–1 is ∅.250 (MMC) and the clearance holes are ∅.280 minimum (MMC). What is the required positional tolerance?

$$T = H - F$$
$$= .280 - .250$$
$$T = .030$$

In this example a positional tolerance of ∅.030 is applied to *each* part. The total tolerance in both parts is .060. If desired, this can be divided unequally between the parts. One might have a tolerance of, say, .040, in which case the tolerance on the other part would be .020.

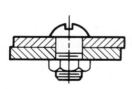 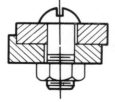 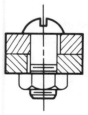

Fig. 18-2 A one-fastener assembly—no clearance required.

Fig. 18-3 One-fastener assemblies—clearance required.

When parts are held together by only one fastener and the parts are positioned only by the holes, no clearance is necessary (see Fig. 18–2). However, if the edges of the parts must be aligned or if one part is located by a shoulder on the other part as in Fig. 18–3, then clearance must be provided to compensate for positional error. This is treated the same as a multiple-hole floating fastener assembly.

18.3 Analysis of Fixed Fasteners

The fixed fastener condition differs from floating fasteners in that the fastener is fixed in one of the parts. It cannot change its location to accommodate positional error. A fastener size is selected and then the hole size and positional tolerance are calculated by the following formulas.

$$H = F + 2T \qquad T = \frac{H - F}{2}$$

In the second formula H – F is equal to the clearance between the hole and the fastener. Thus the positional tolerance is equal to one half of the clearance.

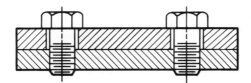

Fig. 18-4 Fixed fastener assembly.

Example: Assume that the fastener in Fig. 18–4 is $\varnothing$.250 (MMC) and the clearance hole in the mating part is $\varnothing$.280 minimum (MMC). What is the required tolerance?

$$T = \frac{H - F}{2} = \frac{.280 - .250}{2}$$
$$T = .015$$

In this example a positional tolerance of $\varnothing$.015 is applied to each part. The total tolerance in both parts is .030. If desired, this can be divided unequally between the parts. One might have a tolerance of, say, .020, in which case the tolerance on the other part would be .010.

As with floating fasteners, when parts are held together by only one fastener and the parts are positioned only by the holes, no clearance is necessary. However, if the edges of the parts must be aligned or if one part is located by a shoulder on the other part, then clearance must be provided to compensate for positional error. This is treated the same as a multiple-hole fixed fastener assembly.

The fixed-fastener formulas do not provide sufficient clearance when tapped holes or holes for press-fit pins are out of square. To provide for this condition the *projected tolerance zone* concept is used. This is explained below.

18.4 Projected Tolerance Zone

With fixed fastener assemblies where the clearance hole is calculated by the fixed-fastener formula, the parts will sometimes not assemble because of positional error that permits the axis of the fixed hole to be out of square with the surface. An example is shown in Fig. 18–5.

Now if the upper part in Fig. 18–5 were very thin, there would be no interference, and if the part were thicker, there would be even more interference.

It is evident, therefore, that the height of the positional tolerance zone for the tapped hole should be based upon the height of the mating part and not on the height of the tapped hole. The projected tolerance zone provides for this by transferring the tolerance zone from the part being controlled to the mating part (see Fig. 18–6).

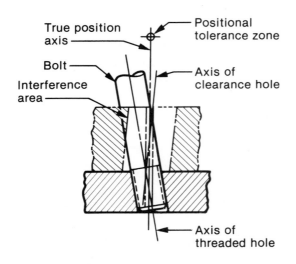

Fig. 18-5   Interference of mating part caused by out-of-squareness of tapped hole axis. (ANSI Y14.5)

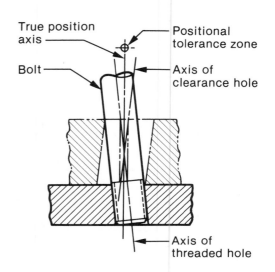

Fig. 18-6   Interference eliminated by making length of positional tolerance equal to height of mating part. (ANSI Y14.5)

Comparing this figure with Fig. 18–5, it is evident that the clearance hole size and the positional tolerance have not been changed. The difference is that in Fig. 18–6 the tolerance zone for the tapped hole is not within the tapped hole but is projected upward to a height equal to the height of the mating part. The mating part, of course, has its own positional tolerance for the clearance hole. The two tolerance zones now coincide and this results in less perpendicularity error. The parts will now assemble properly when normal fixed-fastener clearance hole sizes are used.

An example of a projected tolerance zone specification is given in Fig. 18–7. The drawing of the mating part containing the clearance holes will specify the same positional tolerance without a projected tolerance zone.

Where the fasteners are studs or press-fit pins, the projected tolerance zone must be equal not to the thickness of the mating part but to the installed height of the stud or pin above the surface (see Fig. 18–8).

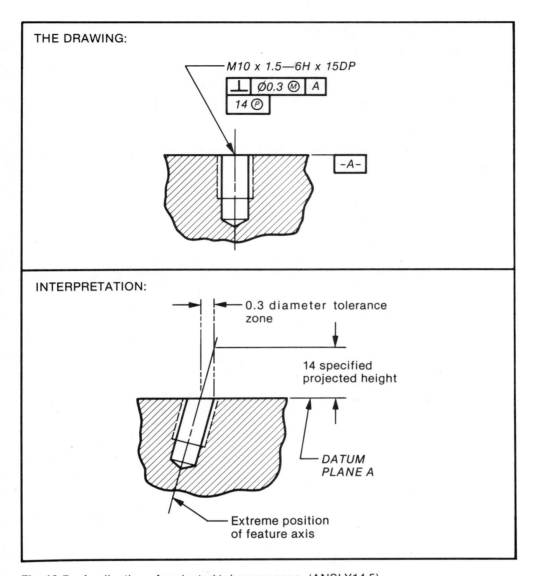

Fig. 18-7 Application of projected tolerance zone. (ANSI Y14.5)

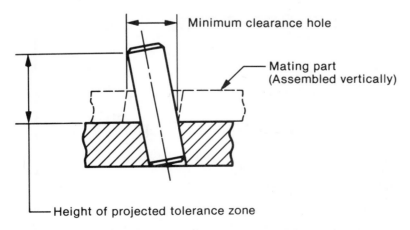

Fig. 18-8 Height of projected tolerance zone.

18.5   Zero Positional Tolerance at MMC

It sometimes occurs that a part may be rejected because a feature exceeds the size limits even though it might actually fit the mating part. An example of this is shown in Fig. 18–9. The minimum size (MMC) is given as $\varnothing$14.25 mm to clear a 14 mm screw. If the produced size were to be, say, $\varnothing$14.10 mm (0.10 mm over the maximum screw size) and the actual positional error were under 0.10 mm, the screw would fit in the hole. However, the part, although usable, would have to be rejected because of the undersized hole. The unused positional tolerance cannot be added to the size tolerance.

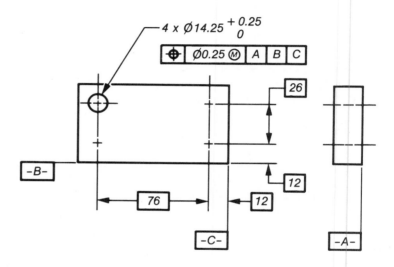

Fig. 18-9   A normal application of positional tolerance at MMC.

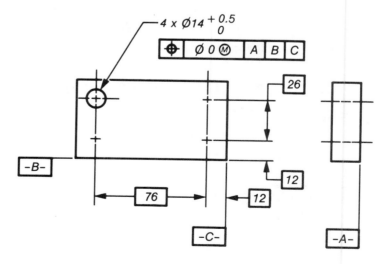

Fig. 18-10   Application of zero tolerance at MMC—same part.

A tolerancing technique has been developed to prevent this. It consists of specifying zero positional tolerance and expanding the size tolerance on the hole. The minimum hole limit is made the same as the maximum mating part or very slightly larger. This is illustrated in Fig. 18–10 where the same part as in Fig. 18–9 is shown with a different tolerance for the holes. Now all the tolerance, which has been increased to 0.5 mm, is in the hole size. The positional tolerance is zero, but that only applies when the holes as produced are minimum (MMC), which is unlikely. The actual positional tolerance on the produced part will be equal to the amount the actual hole size exceeds the MMC, which can be 0.5 mm, a practical amount. There will be no unused positional tolerance. The table below shows the positional tolerance available for various produced hole sizes.

Produced Hole Size (mm)	Clearance with Screw (mm)	Allowable Positional Tolerance (mm)
⌀14	0	0
⌀14.1	0.1	0.1
⌀14.2	0.2	0.2
⌀14.3	0.3	0.3
⌀14.4	0.4	0.4
⌀14.5	0.5	0.5

A slightly larger drill or punch can be used by the manufacturing department (but within tolerance) to produce the holes in Fig. 18–10 to ensure that some positional tolerance will always be available.

In many applications zero positional tolerance at MMC is not suitable, such as when a specific running or sliding fit is required between mating parts.

An often-stated disadvantage of zero positional tolerance is that if both mating parts should be produced at their MMC sizes, the position and form would have to be perfect. But this is a very remote possibility. There is sometimes a psychological problem: The zero tolerance might give an unjustified impression of superior precision and high cost. This problem is generally overcome as personnel receive training and practical experience in zero positional tolerancing.

18.6 Nonparallel Holes

Holes that are not parallel and not drilled or punched perpendicular to a flat surface may also be controlled by positional tolerance. An example is shown in Fig. 18–11. Here, radial holes have been drilled in the wall of a tubular part. The location dimension may be given with a tolerance, as in the example, or it may be a basic dimension.

In the illustration, the location dimension has a tolerance of ±.020, which means that the actual center plane of the four holes may be as much as .020 on either side of the desired center plane. Each of the holes must be within a .010-diameter tolerance zone centered on the same center plane somewhere within that .040 range. This is shown in the interpretation portion of Fig. 18–11.

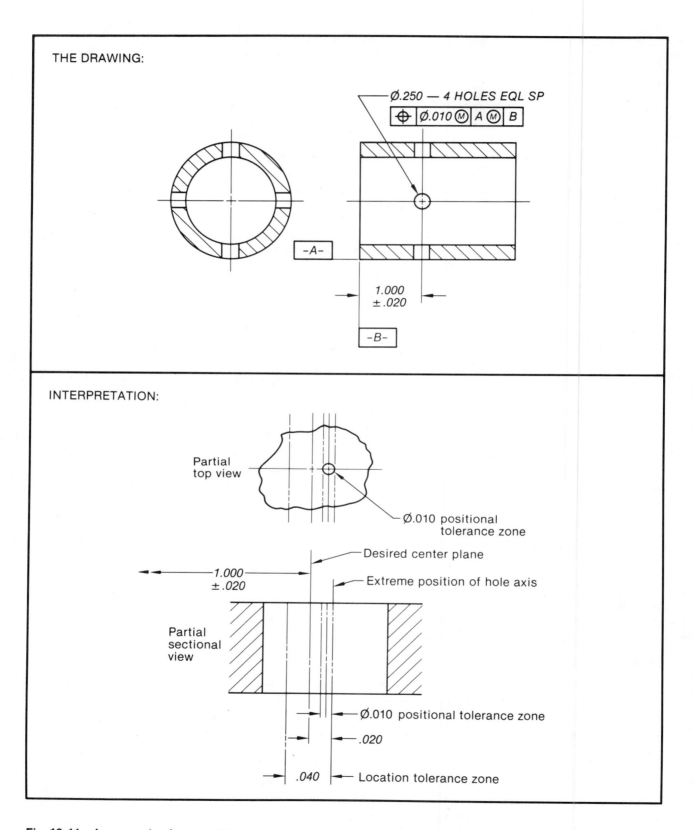

Fig. 18-11   An example of nonparallel holes controlled by positional tolerance.

18.7 Noncircular Features

The benefits of positional tolerance for circular features apply equally well to noncircular features such as slots and tabs. Features of this kind are located not on an axis but on a center plane. It will be recalled from Section 17.3 that for features located on a center plane the form of the tolerance zone is the space between two imaginary planes perfectly parallel to and equidistant from the center plane (see Fig. 17–2). The distance between the imaginary planes is equal to the specified tolerance. The length of the tolerance zone is equal to the length of the feature. The actual center plane of the feature as produced must lie between the imaginary planes.

Fig. 18–12 shows mating parts with slots and tabs controlled by positional tolerances. The tolerances were determined by this formula.

$$T_{tot} = W_s - W_t$$

Where: T_{tot} = Total tolerance in both parts.

W_s = Width of slot at MMC.

W_t = Width of tab at MMC.

Notice that since $W_s - W_t$ equals the minimum clearance between the slot and tab, the total tolerance equals the minimum clearance. This is similar to the floating fastener situation (Section 18.2) except that only two parts are involved (no fasteners).

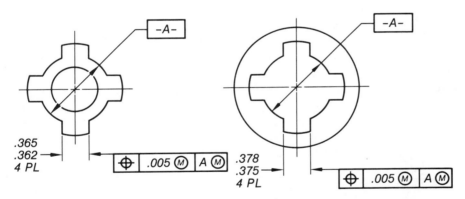

Note: The positional tolerance in both drawings is actually controlling symmetry of the slots or tabs relative to the axis of DATUM A.

Fig. 18-12   Mating parts with noncircular features controlled by positional tolerance.

In the example in Fig. 18–12, W_s = .375 and W_t = .365.

$$T_{tot} = .375 - .365 = .010$$

The total tolerance in this case is split equally between the two parts, each one having a tolerance of .005. If desired, the tolerance can be split in any other combination (.007 and .003, .006 and .004, and so forth). This may be done when a close tolerance is more difficult to hold in one of the parts.

Fig. 18–13 illustrates a typical tab and slot of each part assembled in the virtual (extreme) condition. The tab is the widest possible size and slanted at the extreme angle permitted by the geometric tolerance zone. The slot is its narrowest possible size and slanted to the extreme in the opposite direction. This is the tightest possible fit of the two parts.

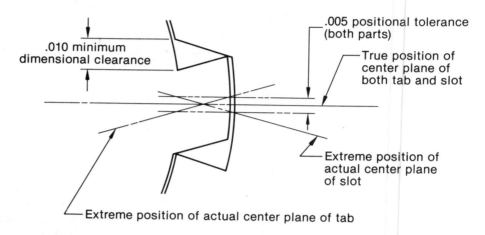

Fig. 18-13   Tightest fit of slot and tab shown in Fig. 18-12.

18.8   Use of Positional Tolerance for Angularity at MMC

In Chapter 14, "Angularity," it was said that although angularity normally applies RFS, it can be specified MMC when size features are involved. However, this is more conveniently done by controlling the feature and/or the datum by positional tolerance. An example is shown in Fig. 18–14. Here both datums and the controlled keyway apply at MMC.

Parallelism and perpendicularity at MMC may also be specified by a positional tolerance, but this is not often done in American engineering practice.

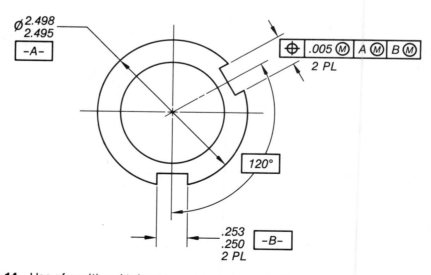

Fig. 18-14   Use of positional tolerance to control angularity.

18.9   Use of Positional Tolerance for Symmetry

Symmetry is the quality of being the same on both sides in size, shape, and relative position from a center plane. The human body is symmetrical on the left and right sides of an imaginary vertical center plane (not front and rear, not top and bottom).

Symmetry applies only to size features. The datum must also be a size feature. *Symmetry error* can be thought of as the amount by which opposite sides of a size feature are unequally spaced from the center plane of the datum. Symmetry tolerance, then, is the total permissible error in symmetry.

The form of the tolerance zone is the same as for any feature located on a center plane, with a slight difference. It is the space between two imaginary planes perfectly parallel to and equidistant from the true position of the center plane *of the datum* (see Fig. 17–2).

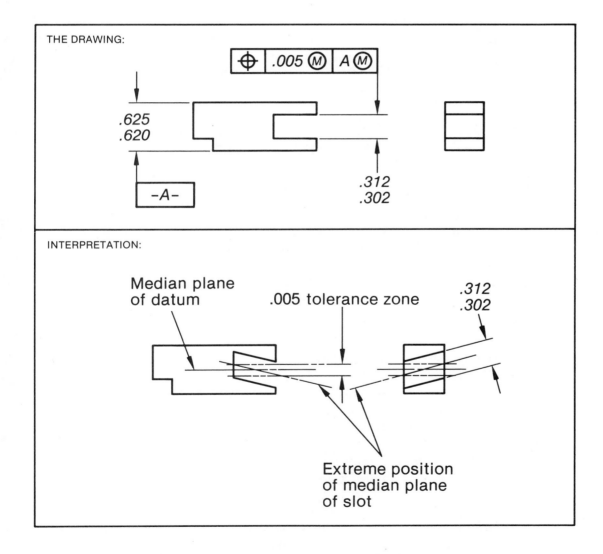

Fig. 18-15   Use of positional tolerance to control symmetry of a slot.

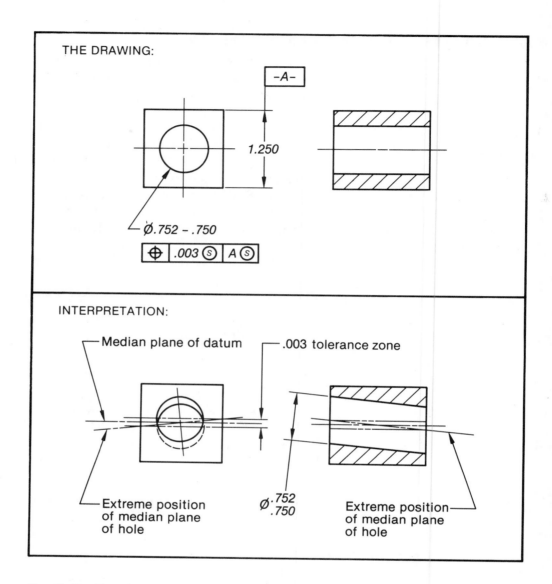

Fig. 18-16 Use of positional tolerance to control symmetry of a hole.

Symmetry is actually a special case of positional tolerance, and this indeed is how it is treated and specified on drawings. Until recently a special geometric characteristic symbol was used, but, starting with ANSI Y14.5M, positional tolerance was expanded to include symmetry. Fig. 18–15 and Fig. 18–16 are examples. The two drawings in Fig. 18–12, which were studied as location tolerances, are also symmetry applications.

Depending upon the design requirements, a symmetry tolerance may apply at MMC as in Fig. 18–12 and Fig. 18–15, or RFS as in Fig. 18–16, or it may apply at LMC.

It is very difficult to measure from a center plane—since it is imaginary—therefore, symmetry error is measured to the edge of the feature, as shown in Fig. 18–17. A measurement is taken from one side of the feature to one side of the datum. Then the part is rotated 180° and another measurement is taken from the other side of the feature to the other side of the datum. The difference between the two measurements is equal to the error in symmetry.

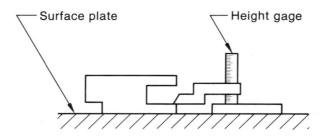

First Measurement

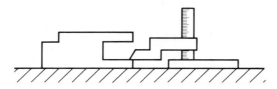

Second Measurement.   Part rotated 180°

Typical readings
   First: .153
   Second: .157
Symmetry error = .157 − .153 = .004

Question: Should part be accepted or rejected?

Fig. 18-17   Typical symmetry inspection of part shown in Fig. 18-15.

Zero positional tolerance at MMC may be applied to symmetry as easily as to other types of positional tolerance. Taking the part in Fig. 18–15 as an example, the positional tolerance could be reduced to .000 at MMC. The actual permissible positional error would then be equal to the amount that the feature and datum sizes as produced depart from MMC. The maximum would be the sum of the size tolerances (.005 + .010 = .015). In the unlikely event that the datum were produced .625 and the slot were produced .302 (both at MMC), the symmetry would have to be perfect.

18.10   Multiple Patterns of Features

Patterns of features are any arrangement of features on an object. The features are usually circular holes, but they may be any kind of opening (round-end or rectangular slots, or any irregular shape), or they may be protrusions such as bosses, pads, or embossed figures, or they may be simply markings on the surface. Although all of the examples following will show *hole* patterns, the information applies equally well to all types of feature patterns. When there are two or more feature patterns on the same part, they are called *multiple* patterns of features.

THE DRAWING:

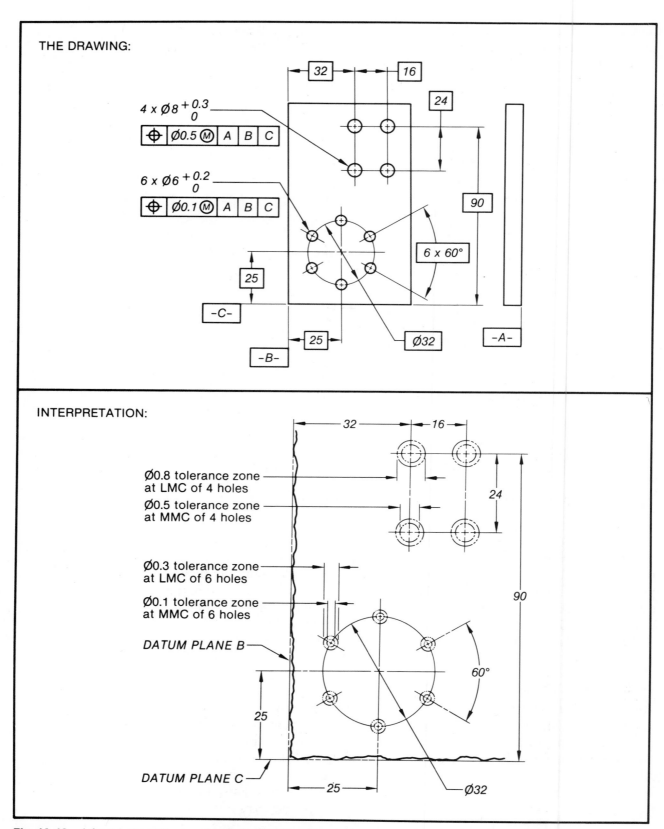

INTERPRETATION:

Fig. 18-18 A four-hole pattern and a six-hole pattern treated as one composite pattern. (ANSI Y14.5)

Multiple patterns of features located by basic dimensions from the same datums (but *not* size datums) are considered one composite pattern. Fig. 18–18 is an example. The four-hole rectangular pattern and the six-hole circular pattern will be inspected with one gage, as if they were one ten-hole pattern. If, because of the complexity of the part or its size, it is necessary to use two gages, the two hole patterns are still treated as one composite pattern. Both gages must seat against datums A, B, and C.

Multiple patterns of features located by basic dimensions from the same datums that are *size* datums are also considered one composite pattern when their respective feature control frames are identical except for the tolerances. The tolerances may be the same or different. The datums must be specified in the same order and the modifiers must agree for both or all patterns. In Fig. 18–19 the positional tolerances for the two 2-hole patterns are different, but everything else in the feature control frames is identical; therefore, the two 2-hole patterns are treated as one composite pattern.

In both situations above, if it is required for any reason that the hole patterns be treated as separate requirements and gaged separately, a note "SEP REQT" (separate requirement) is drawn below each feature control frame. This might be done, for example, if the two small holes in Fig. 18–19 are used for attaching another part and the two large holes are used for mounting onto the next assembly.

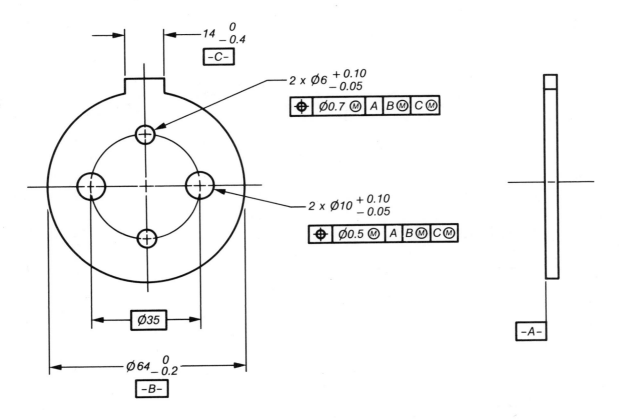

Fig. 18-19 A part with two two-hole patterns located from the same datums, two of which are size features. (ANSI Y14.5)

Question: Are the two patterns treated as one composite pattern?

18.11   Composite Tolerances for Feature Patterns and Individual Features Within the Patterns

It often happens that the positional location tolerance for a pattern of holes or other features should be different (usually larger) than the positional tolerance of the individual holes. In this case two feature control frames are drawn as in Fig. 18–20, one below the other, sharing one positional tolerance symbol. The upper frame always gives the tolerance for the location of the pattern, and the lower one for the location of individual holes within the pattern.

In this case three datums are used for the pattern location, of which A is the first datum (the most important), since the part will be seated on that surface in the next assembly. Datums B and C are the second and third datums, respectively. They locate the hole patterns dimensionally and provide the required orientation on the part, both within a $\varnothing 0.8$ mm tolerance zone. The individual holes are located within each pattern by basic dimensions. Datum A is specified to control

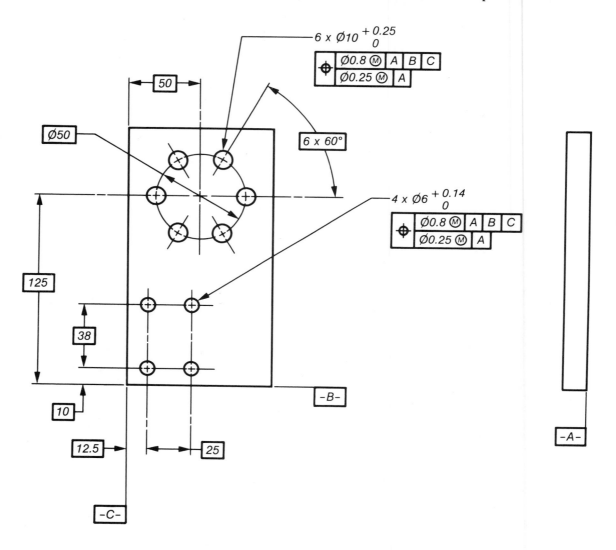

Fig. 18-20   A plate containing two hole patterns and composite positional tolerances. (ANSI Y14.5)

perpendicularity within the ⌀0.25 mm tolerance zone. An interpretation is given in Fig. 18–21 for the circular pattern of holes. The permissible positional variation for the pattern consists of six ⌀0.8 mm tolerance zones perfectly centered in their true position. The center of each ⌀0.25 mm tolerance zone for the individual holes must be within the ⌀0.8 mm tolerance zone and located on the intersection of six equally spaced radial lines and a ⌀50 mm circle. The actual axis of each hole must be somewhere within the ⌀0.25 tolerance zone and also within the larger tolerance zone. Note in Fig. 18–21 that the radial lines and circle on which the smaller tolerance zones are centered may be slightly removed from the desired location (shown on the drawing) on which the larger tolerance zones are centered.

The interpretation is shown with the holes at their MMC size (smallest diameter). When the holes are at their LMC size, the positional tolerance of the pattern increases to ⌀1.05 mm and that of the individual holes increases to ⌀0.50 mm.

On many existing drawings, feature patterns are located by dimensions with plus-and-minus tolerances while the individual features are controlled by basic dimensions and positional tolerances. This practice is not advocated by ANSI Y14.5M and will not be analyzed in this text (see Appendix A).

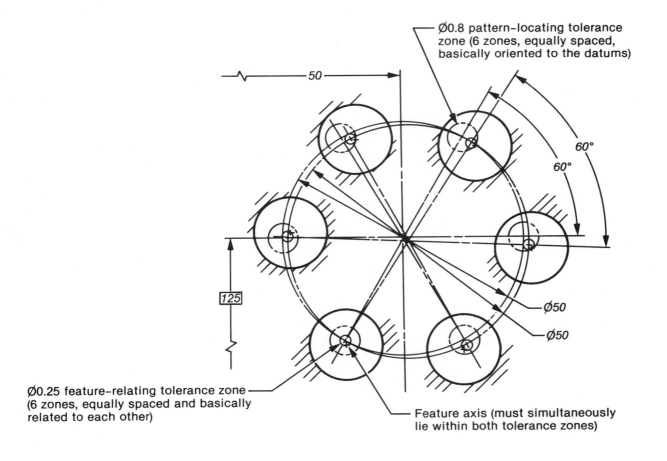

Fig. 18-21   Interpretation of positional tolerance for the circular hole pattern in Fig. 18-20. (ANSI Y14.5)

CHAPTER 19

POSITIONAL TOLERANCE— COAXIAL APPLICATIONS

19.1 Definition

Two or more geometric figures are said to be coaxial if they have the same axis or if their axes are in line (Section 15.1). Any two *regular* shapes may be coaxial. This includes figures such as cylinders, cones, curved profiles, and square or hexagonal shapes, all of which are illustrated in Fig. 15–1. This study will be based mostly on cylinders, but all of the information applies equally well to other regular shapes.

In discussing coaxiality in chapters 15 and 16, it was said that there is no symbol for the general case of coaxiality; but there are three symbols that define special cases. These are concentricity, runout, and positional tolerance.

19.2 Form of the Tolerance Zone

The form of the tolerance zone for coaxial positional tolerance is an imaginary cylinder perfectly coaxial with the datum axis. The diameter of the cylinder is equal to the specified tolerance; its length is equal to the length of the feature (see Fig. 19–1). This is the same form as for any feature located by its axis for any type of geometric tolerance. The tolerance is always specified as a diameter ($\varnothing$) in the feature control frame.

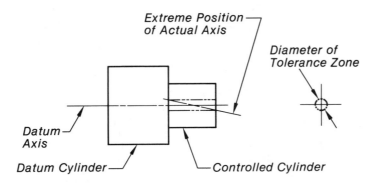

Fig. 19-1 The form of the tolerance zone for coaxial positional tolerance.

19.3   Surface and Orientation Errors

Coaxial positional tolerance is a composite tolerance. In addition to coaxiality it includes errors on the surface of the feature, such as roundness and straightness. It also includes orientation errors; that is, the axis of the feature might be slanted relative to the datum. This is similar to runout, also a composite tolerance, but different from concentricity, which is pure coaxiality.

19.4   Use of Modifiers

In recent years, positional tolerancing has become the preferred method of controlling coaxiality for most applications because, unlike concentricity, it is easy to measure, and the tolerance can be (and almost always is) specified at the MMC size of the feature, which provides the designer with the much desired *bonus* tolerance discussed in Section 17.7. Runout, which also controls coaxiality, is always applied RFS. A comparison of positional tolerance, runout, and concentricity is made at the end of this chapter.

MMC is specified for the feature and for a size datum whenever possible so that additional tolerance may be permitted when the size as produced is not at MMC. When the reason for specifying coaxiality is "assemblability" (so that parts will assemble properly), the MMC modifier is appropriate. The "circle M" symbol Ⓜ is drawn in the feature control frame after the tolerance and, when desired, after the datum letter. Examples are shown in Fig. 19–2.

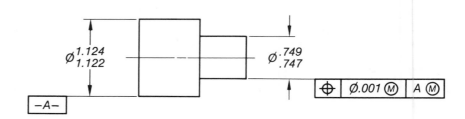

Fig. 19-2   Coaxial positional tolerance at MMC.

LMC is not often specified in coaxiality applications. In situations where a wall thickness is more critical than coaxiality, LMC could be used to maintain a certain minimum thickness. In the part shown in Fig. 19–3, the minimum wall thickness, considering the size tolerances and the positional tolerance at LMC, is 1.85 mm.

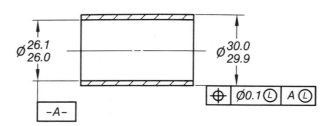

Fig. 19-3   Coaxial positional tolerance at LMC.

As the produced sizes of the O.D. and I.D. depart from LMC, more positional tolerance is permitted. This is shown for three possible sizes in the table below.

Produced Size (mm)		Total Departure from LMC (mm)	Permissible Positional Error (mm)	Min. Wall Thickness (mm)
I.D.	O.D.			
26.1	29.9	0	0.1	1.85
26.05	29.95	0.1	0.2	1.85
26.0	30.0	0.2	0.3	1.85

The total departure from LMC for both the I.D. and O.D. is 0.2 mm, which can be added to the positional tolerance as a *bonus tolerance* with no decrease in the critical wall thickness.

Coaxial positional tolerance can be applied RFS simply by adding "circle S" (Ⓢ) after the tolerance and/or the datum letter. In Fig. 19–4 the positional tolerance on the smaller cylinder applies at MMC but Datum C is RFS. In coaxiality situations where features and/or datums are to apply RFS, it is common practice to control the coaxiality by runout rather than positional tolerance. Runout automatically applies RFS. No modifier is needed.

19.5   Specifying Coaxial Positional Tolerance

Examples of drawings in which coaxiality is controlled by positional tolerance are given in Figs. 19–2 through 19–7.

Fig. 19–4 shows a cylinder (C) coaxial with a rectangular block in the middle of the part. Both the cylinder and the block apply at MMC. The smaller cylinder on the left is coaxial with Cylinder C. Note that in the feature control frame for the smaller cylinder the positional tolerance applies at the MMC (∅.499) of the smaller cylinder but regardless of the size of Datum C, the larger cylinder.

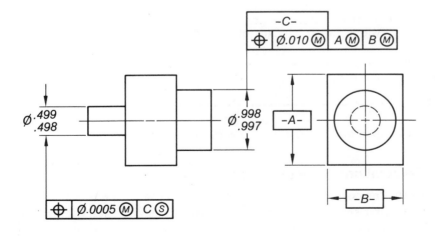

Fig. 19-4   Coaxiality between a rectangular block and two cylinders—MMC and RFS specified.

19.6   Coaxial Holes

The alignment of two or more holes on a common axis may be controlled by coaxial positional tolerance. It is used where a location tolerance alone does not provide the necessary control of coaxiality. Fig. 19–5 shows an example of four coaxial holes of the same diameter. The axes of all four holes must be within a common ∅0.15 tolerance zone (coaxiality tolerance). At the same time the ∅0.15 tolerance zone must be somewhere within a ∅0.25 tolerance zone (location tolerance) centered on the true position relative to datums A and B.

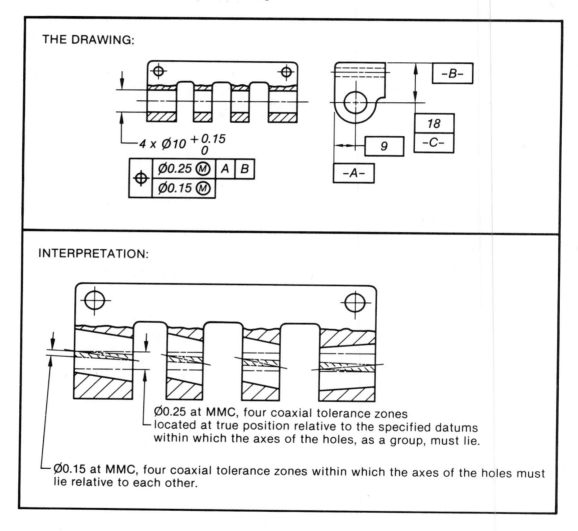

Fig. 19-5   Control of coaxial holes by positional tolerance. (ANSI Y14.5)

19.7   Calculations to Determine Coaxial Tolerance

To determine the correct positional tolerances to use with mating coaxial parts, the designer first selects the dimensional tolerances for each diameter of both parts. This will result in certain maximum and minimum clearances between the parts. To avoid interference the total positional tolerance for both parts must not be greater than the minimum clearance.

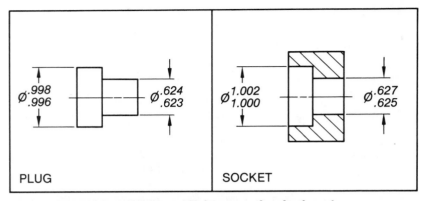

(a) Drawings Before Positional Tolerances Are Assigned

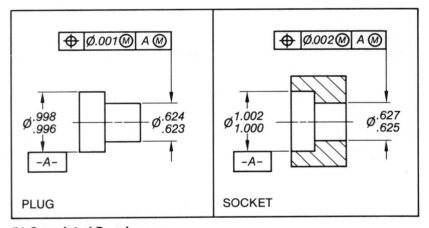

(b) Completed Drawings

Fig. 19-6   Mating plug and socket with calculated
positional tolerances to ensure desired fit.

Fig. 19–6 will be used as an example. Drawings of two mating coaxial parts are shown. The plug is to fit into the socket at both diameters with a metal-to-metal fit at the tightest condition. The loosest condition will be determined by the sum of the size tolerances on mating dimensions. In part (a) of the illustration the drawings are shown with dimensional size tolerances already assigned but with no positional tolerances. The following three-step procedure will be used to obtain appropriate coaxial positional tolerances for both parts. (In the general case the opening in one of the parts is referred to as "the hole," and the part that fills the opening is called "the shaft.")

1. Obtain the minimum clearance between each set of mating diameters on the two parts.

Socket, minimum hole	.625	1.000
Plug, maximum shaft	−.624	− .998
Minimum clearance	.001	.002

2.   Add the minimum clearances. The sum is the total positional error that can be allowed for *both* parts.

.001 + .002 = .003

3.   Divide the total positional error between the two parts, not necessarily into equal shares.

Of the .003 total tolerance, assign .001 to the shaft (plug) and .002 to the hole (socket).

More tolerance is generally allowed for the hole because coaxiality is more difficult to maintain on internal operations.

Although the minimum clearance between the two parts is .001 (at the small diameters) the plug and socket will actually make line contact when they are both at their MMC (largest shaft; smallest hole) because of the out-of-coaxiality allowed by the positional tolerances.

Part (b) of Fig. 19–6 shows the completed drawing with positional tolerances added. The smaller diameter is the controlled feature in both drawings and the larger diameter is the datum, but this could have been reversed with no effect on the fit. Both the controlled features and the datums apply at their MMC, so that as the produced sizes depart from MMC more positional tolerance is allowed. The tables below show the effect of certain produced sizes on the positional tolerances. Not all possible combinations of controlled feature size and datum size are shown.

PLUG

Controlled Feature	Datum	Total Departure from MMC (Bonus Tolerance)	Permissible Positional Error
∅.624	∅.998	0	∅.001
∅.6235	∅.997	.0015	∅.0025
∅.623	∅.996	.003	∅.004

SOCKET

Controlled Feature	Datum	Total Departure from MMC (Bonus Tolerance)	Permissible Positional Error
∅.625	∅1.000	0	∅.002
∅.626	∅1.001	.002	∅.004
∅.627	∅1.002	.004	∅.006

So, it is seen that although the drawings specify positional tolerances of .001 and .002, the inspector can accept as much as .004 and .006, respectively, without sacrificing proper fit. Since most parts are produced somewhere between the high and low size limits, the probable positional tolerances permissible will be between the extremes—say, .002 for the plug and .003 for the socket.

19.8   Zero Tolerance at MMC

Zero tolerance at MMC can be used as effectively with coaxial features as with other types of positional situations. In Fig. 19–7 is a drawing of the same part as in Fig. 19–2. Here the small cylinder must be perfectly coaxial (zero positional tolerance) with the datum cylinder when both features are at MMC. The size tolerances on both features, however, allow a total variation of .004, and this can be used as positional tolerance. If both cylinders were produced at their *least* material condition (∅.747 and ∅1.122), the coaxiality error could be a maximum of ∅.004. The possibility that both features will be produced at their MMC sizes is very remote. In all likelihood they will each be about .001 under the maximum limit, providing a coaxiality tolerance of about ∅.002.

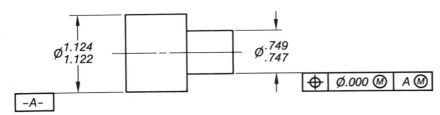

Fig. 19-7   Specifying zero tolerance at MMC.

As is seen in Section 18.5, zero positional tolerancing is a practical alternate method of controlling positional variations while avoiding unused tolerances and possible rejection of usable parts.

19.9   Selection of Proper Control for Coaxiality

Positional Tolerance: Use when

1.   Surface errors may be included with coaxiality error.

     No separate inspection will be made for surface errors unless specified on the drawing.

2.   The coaxial features must assemble with another part having corresponding coaxial features, and additional coaxiality error may be allowed when the feature and/or the datum are not at the maximum material condition (see Section 19.4).

Concentricity: Use when

1.   Coaxiality must be controlled independently of surface errors.

     Separate feature control symbols may be used to control surface errors.

2.   The desired coaxiality tolerance must be held regardless of feature and datum size (RFS).

Runout: Use when

1.   Surface errors may be included with coaxiality error.

     No separate inspection will be made for surface errors unless specified on the drawing.

2.   The desired coaxiality tolerance must be held regardless of feature and datum size (RFS).

CHAPTER 20

THE DATUM FRAME

20.1 Introduction

In cases where surfaces are particularly uneven, such as castings and forgings, or where the part is so thin as to be bowed or warped, it is necessary to specify the desired points of contact between each datum feature and its datum plane. (Datum *feature* and datum *plane* are explained in Section 4.2.) This needs to be done in three planes at right angles to each other since real objects are three-dimensional.

Points of contact may also be specified on machined parts when these points are to serve as the basis for geometric tolerances.

20.2 The Datum Frame

An unsupported object in space can be moved in six directions, technically denoted as six degrees of freedom. This is illustrated in Fig. 20-1. The object can be moved these six ways:

Sliding motion in three directions:

1. Left to right (the X direction)
2. Up and down (the Y direction)
3. Front to rear (the Z direction)

Rotational motion about three axes shown in the figure as:

4. X—X
5. Y—Y
6. Z—Z

Such an object can be accurately located by preventing motion in all of the six degrees of freedom. This is most easily done by supporting it on three surfaces at right angles to each other.

A rectangular object is shown in Fig. 20-1. The specified points where contact is made with the datum planes are called *datum targets* or *tooling points*. Notice that the object has three tooling points on one surface. This is called the first datum. These three points define, in this case, the horizontal plane. The vertical

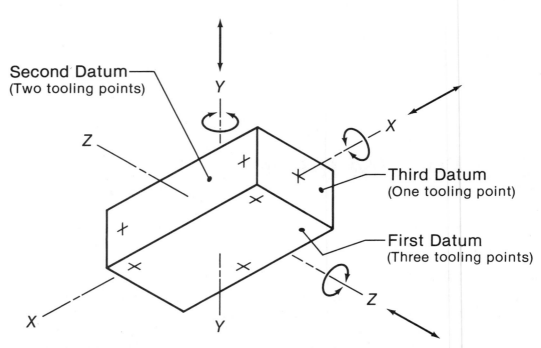

Fig. 20-1   A rectangular object in space, showing the six degrees of freedom and typical tooling points.

plane at the front is the second datum. Only two points are required to define this plane, since the part cannot tip when it is in contact with a horizontal datum plane. Only one point is required in the remaining plane, the third datum, because, with the part in contact with the first and second datums, the only way it can move is to slide horizontally, and one tooling point will prevent that. In actual practice the part is clamped against all three datum planes so it does not move as it is being machined or inspected.

Such a system of three datum planes at right angles is called a *datum frame*.

20.3   Selection of Datums

The designer selects the three datum planes on a particular part based upon the function of the part and its fit in the next assembly. The feature most important to the fit and function is selected as the first datum; the next most important, perpendicular to the first, becomes the second datum; and a remaining mutually perpendicular feature is used as the third datum.

20.4   Specifying Datum Targets on Drawings

Datum targets are represented on the drawing as shown in Fig. 20–2. Each datum target is identified by a *datum target symbol* with a leader pointing to the actual location. The leader has no arrowhead. The symbol is a thick circle divided by a thick horizontal line through the center. The circle diameter is specified in ANSI Y14.5M as 3.5 times the lettering height. For 5/32 (4 mm) lettering this is ∅9/16 (∅14 mm). The datum identifying letter followed by a target number is drawn in the lower half of the circle. Datum target symbols A1, A2, and A3 define the first datum, target symbols B1 and B2 define the second, and C1 defines the third.

The location and size, where applicable, of datum targets are defined with either basic or toleranced dimensions. When basic dimensions are used, tool or gage tolerances apply. Tolerances for tools and gages are normally a small fraction of production tolerances.

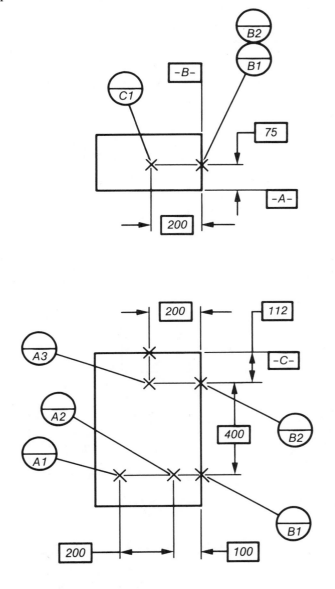

Fig. 20-2 A typical drawing with datum target symbols. (ANSI Y14.5)

Datum targets, although often referred to as tooling *points,* may be in any of three forms: a point, a line, or a small area. A datum target *point* is represented on the drawing with a 45° cross, as shown in Fig. 20-2. The actual tooling may not be a true point, since this would make a mark on the part (brinelling). It is common practice to contact the work with a bullet-nosed dowel to prevent brinelling (see Fig. 20-3).

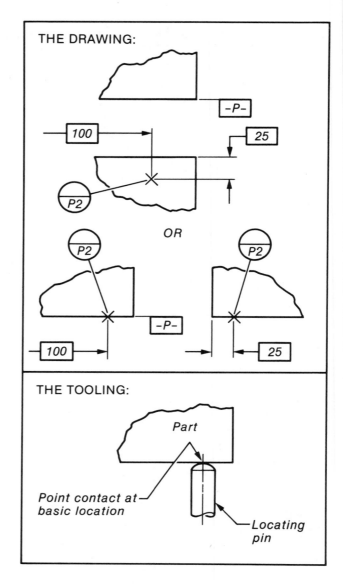

Fig. 20-3 Datum target points and typical tooling. (ANSI Y14.5)

A datum target *line* is represented as a phantom line. In a view where the line would appear as a point, it is shown as a 45° cross (see Fig. 20–4). The illustration also shows that line contact is usually achieved in the tooling by the side of a dowel.

The datum target may also be a small surface (a datum target *area*), which might be a circular area or any other desired shape. If a circular area is used, the diameter is specified in the upper half of the datum target symbol (see Fig. 20–5). The circular area may be shown on the drawing as in (a) or only its position indicated by a cross as in (b). For any shape other than a circle the configuration is drawn and dimensioned on the appropriate view. Phantom lines are used for the outline and the area is cross-hatched (see Fig. 20–7).

THE DRAWING:

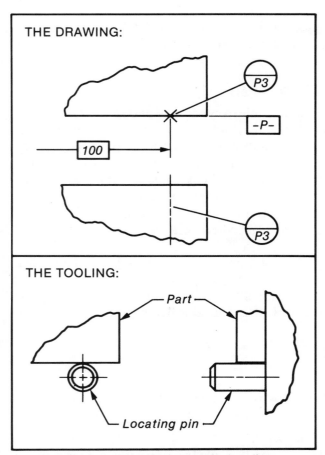

THE TOOLING:

Fig. 20-4   Datum target line and typical tooling. (ANSI Y14.5)

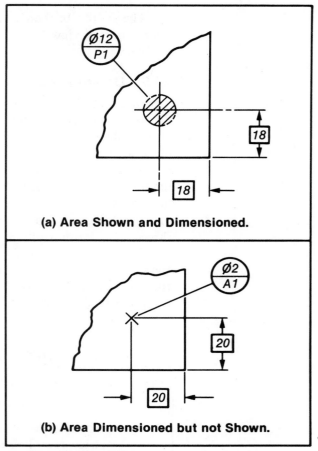

(a) Area Shown and Dimensioned.

(b) Area Dimensioned but not Shown.

Fig. 20-5   Datum target area.

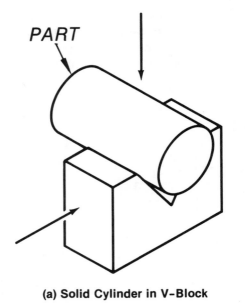

PART

(a) Solid Cylinder in V–Block

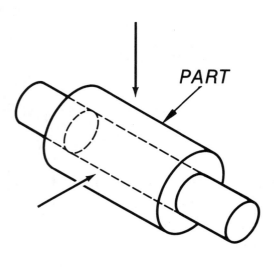

PART

(b) Hollow Cylinder on Mandrel

Fig. 20-6   A cylinder provides location in two directions.

The use of a broken leader line in a view indicates that the datum target is not visible in that view (see Fig. 20–9). Notice that the cross is shown as if it were visible.

20.5   Datum Frame for Cylindrical Parts

The three-plane datum concept is applicable to cylindrical parts even though they do not have three perpendicular planes. A cylindrical surface is equivalent to two datum planes since it automatically establishes location in two perpendicular directions at the same time. In machining and inspecting cylindrical parts, the cylindrical surface may be held in a vee-block, chuck, or collet, or on an expanding mandrel, which restrains movement in two perpendicular directions, usually horizontal and vertical (see Fig. 20–6). If one flat face of the part, or just a point on that face, is used for a third locator, there will be the equivalent of a three-plane datum frame.

When an entire cylinder is used as a datum and no datum targets are specified, the size of the actual datum (supporting tooling) may vary depending on whether the cylinder is internal (an I.D.) or external (an O.D.) and whether the geometric tolerance is specified RFS, MMC, or LMC. The table below presents these data.

SIZE OF DATUM

	RFS	MMC	LMC
I.D.	Largest cylinder that will fit inside, regardless of I.D. size.	Lower limit of I.D. size	Upper limit of I.D. size
O.D.	Smallest cylinder that will fit over the outside, regardless of the O.D. size	Upper limit of O.D. size	Lower limit of O.D. size

Datum targets may be specified on cylindrical surfaces when necessary. An example might be a shaft supported by bearings as in Fig. 20–7. The datum target locations duplicate the positions of the bearings. Datum target A1 is an area, datum target A2 is a line (a circle). An area here would be too restrictive since the shaft is already restrained over an area at A1. During inspection of this part a stop

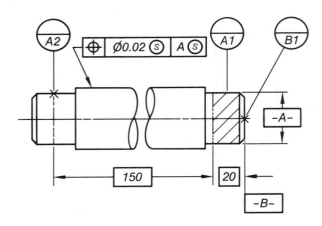

Fig. 20-7   Datum target area and line applied to a
cylindrical part (metric). (ANSI Y14.5)

will be placed against datum target B1 to prevent axial motion while the part is rotated.

Another example of datum targets on cylindrical parts is shown in Fig. 20–8. Here three points each are designated on as-cast surfaces A and B. Cylindrical Datum B will be established by a contacting device having three equally spaced simultaneously adjustable radial pins located about an axis perpendicular to primary Datum A.

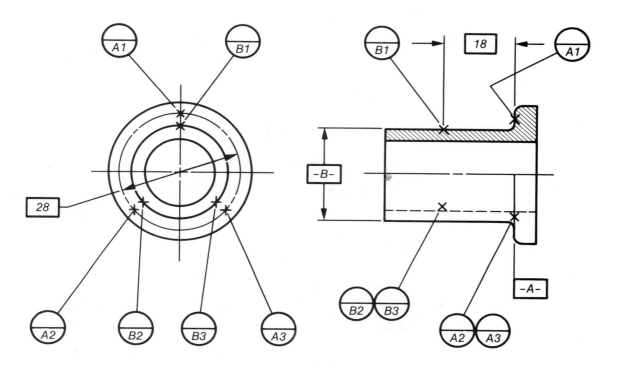

Fig. 20-8   Datum targets (tooling points) applied to a cylindrical machined casting (metric).

20.6   Step Datums

A datum plane may have an offset or step in it as shown in Fig. 20–9. Datum A in this part has two tooling points on the bottom of the large boss and one other on the bottom of the small boss, 20 mm higher. Note that where a basic dimension is used to locate the offset tooling point a positional tolerance is required for the part itself. An alternate practice is to add a duplicate location dimension with a tolerance (say, 20 ± 0.8).

20.7   Equalizing Datums

Round-end parts such as links and connecting rods are supported for machining and inspection on their flat surfaces and located by their round ends. A typical round-end part is shown in Fig. 20–9, where two datum target lines are specified at the large end and two datum target points are specified at the small end. The line datum, Datum B, can be achieved by a fixed line-contact vee-block. Datum C may be established by two bullet-nosed pins that are made to move radially

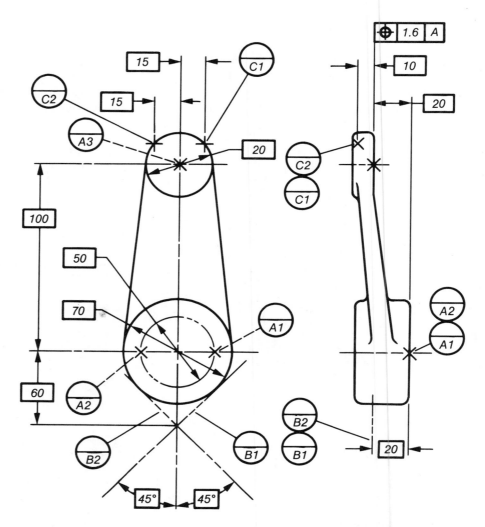

Fig. 20-9 Application of step datum, equalizing datum, and invisible datum targets. (ANSI Y14.5)

inward equally. As one moves a given distance to contact the periphery of the part, the other moves the same distance, thus centering the round end between them. These are called *equalizing pins*. A datum established by any such equalizing method is designated as an *equalizing datum*.

SUMMARY OF NONSTANDARD PRACTICES

A.1 Introduction

Thousands of drawings are still in use that treat geometric tolerancing in different ways from the methods specified in ANSI Y14.5M and explained in this text. Also, many design groups all over the country are not yet using the current ANSI standard. For these reasons, this appendix has been added to explain nonstandard practices that are still being applied to current drawings and that will also be found on older drawings still in use for production, rework, or spare parts. The intent of these geometric specifications may be explained in the title block, the general notes, the company Drafting Room Manual (DRM), or in a manufacturing standard. Or the intent may have been only in the mind of the designer.

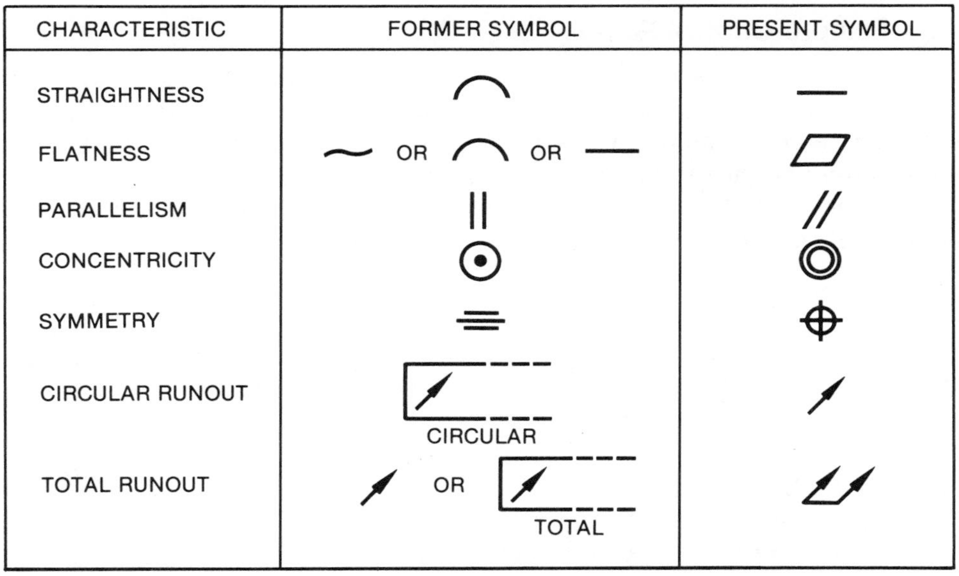

CHARACTERISTIC	FORMER SYMBOL	PRESENT SYMBOL
STRAIGHTNESS	⌒	—
FLATNESS	～ OR ⌒ OR —	▱
PARALLELISM	‖	//
CONCENTRICITY	⊙	◎
SYMMETRY	≡	⨁
CIRCULAR RUNOUT	↗ CIRCULAR	↗
TOTAL RUNOUT	↗ OR ↗ TOTAL	↗↗

Fig. A-1 Formerly used symbols.

A.2 Formerly Used Symbols

The symbols for straightness and flatness have undergone some evolution as is seen in Fig. A–1. The older symbols were used in MIL-STD-8, which was the standard on dimensioning and tolerancing used by United States government agencies until 1966, when the government adopted ANSI Y14.5.

The older vertical-line parallelism symbol had exactly the same meaning as the modern symbol.

The concentricity symbol on many drawings was intended to denote coaxiality including surface errors, which is actually what is now called runout. When such a drawing is used and it is known that runout was actually intended, the drawing should be treated as though the geometric characteristic were runout. If the intent is not known, the concentricity symbol must be taken literally and the part produced and inspected for true concentricity. An optional form of the concentricity symbol had a filled-in inner circle. This had no effect on the meaning.

The older symmetry symbol was abandoned in favor of the positional tolerance symbol by ANSI Y14.5M. When the symmetry symbol was used, the tolerance automatically applied RFS unless MMC or LMC was specified. The meaning was otherwise identical to positional tolerance as explained in Section 18.9. When positional tolerance is specified, a modifier must be applied to the tolerance and to the datum if it is a size feature.

There has been considerable confusion regarding runout. The symbol ∕, which is now standardized as circular runout, *has* been used to denote total runout. If circular runout was in fact intended the word CIRCULAR was added below the feature control frame. The same symbol at another time and in other places was used in the opposite way. In the 1973 edition of ANSI Y14.5 the symbol ∕, unless otherwise specified, denoted circular runout, as it does now. Total runout was specified by adding the note TOTAL below the feature control frame.

A.3 Feature Control Frame

The feature control frame was formerly called the feature control *symbol* and has passed through other phases before acquiring its present configuration. Originally, the frame had no partitions and the sequence of entries was different—the datum letter and the tolerance were reversed. The abbreviation DIA was used in place of the symbol ∅ [see Fig. A–2 (a)]. Later, partitions were added but the sequence of entries remained the same (reversed from present usage), as shown in Fig. A–2 (b).

(a)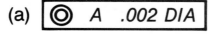

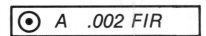

(b)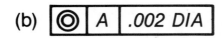

Fig. A-2 Former configurations of the feature control frame.

Fig. A-3 Former designation of diametral tolerance.

At one time tolerances for geometric characteristics that can be inspected by the use of a dial indicator were labelled TIR (total indicator reading) or FIR (full indicator reading) (see Fig. A–3). The abbreviations TIR and FIR were identical in meaning to the currently standard FIM (full indicator movement).

A.4 Size of the Tolerance Zone

There has been a variation in the concept of the size of the tolerance zone. In modern practice the tolerance given in the feature control frame is the *total* width or the *diameter* of the tolerance zone. Some designers in the past thought of the permissible error as the deviation from perfect geometry in *either* direction, so the tolerance given was *half* the total width for a rectangular tolerance zone. Sometimes "R" was specified with a cylindrical tolerance but there was no label indicating that a rectangular tolerance was half the total width. When a tolerance was meant to apply to the total width or the diameter, the word TOTAL or the abbreviation DIA was added after the tolerance value. TOTAL was often abbreviated TOT.

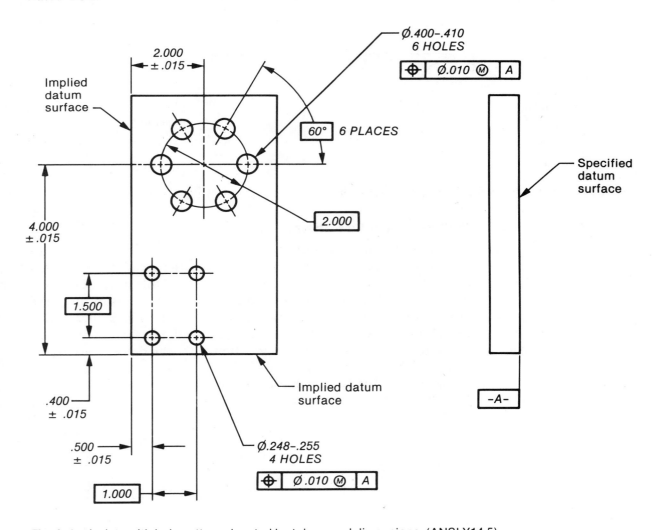

Fig. A-4 A plate with hole patterns located by toleranced dimensions. (ANSI Y14.5)

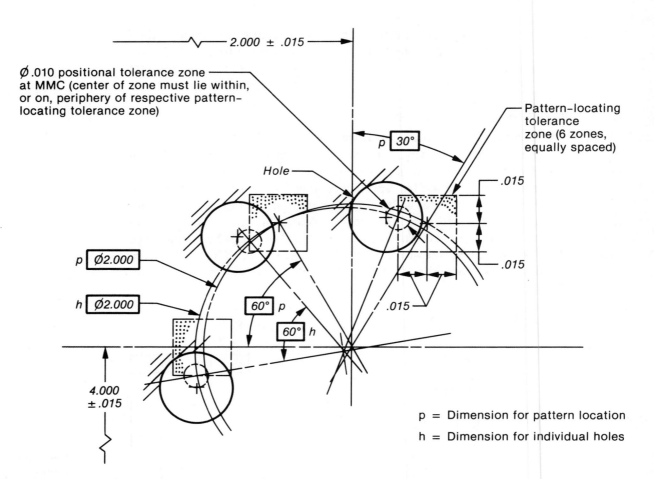

$\emptyset$ .010 positional tolerance zone
at MMC (center of zone must lie within,
or on, periphery of respective pattern-
locating tolerance zone)

2.000 ± .015

Pattern-locating
tolerance
zone (6 zones,
equally spaced)

p 30°

Hole

p $\emptyset$2.000

h $\emptyset$2.000

60° p

60° h

.015

.015

.015

4.000
±.015

p = Dimension for pattern location

h = Dimension for individual holes

Fig. A-5   Interpretation of the tolerance zones for the circular hole pattern in Fig. A-4.

A.5 Location of Feature Patterns

On many existing drawings, feature patterns are located by dimensions with plus-and-minus tolerances from unlabelled (implied) datums while the individual features are controlled by basic dimensions and positional tolerances. The standard method is to use basic dimensions and positional tolerances for both the pattern location and individual feature location, as explained in Section 18.11. An example of hole patterns located by dimensions with plus-and-minus tolerances is shown in Fig. A–4, and the interpretation of the tolerance zones for the circular pattern of holes is given in Fig. A–5. This part is similar to that in Fig. 18–20 and Fig. 18–21. The fault with toleranced coordinate dimensions is that they produce a square tolerance zone wherein the deviation from the true position (the center of the square) is not the same in all directions (see Section 17.4).

CONVERSION TABLES—
COORDINATE TOLERANCE
TO POSITIONAL TOLERANCE

B.1 Introduction

It is often necessary on the job to convert coordinate tolerances to equivalent positional tolerances. This appendix provides two tables for that purpose. Table B-1 will be used mostly by engineering personnel to obtain tolerances to specify on drawings. Table B-2 will find more frequent use by manufacturing and quality control personnel to compare positional tolerances on prints with measured quantities on the hardware.

B.2 Conversion of Bilateral Tolerance to Positional Tolerance

In Table B-1, bilateral (plus or minus) tolerances are given on a horizontal and vertical grid. Positional tolerances are represented by the 90° arcs.

Example: Convert a drawing tolerance of ±.0035 which locates a feature horizontally (the X axis) and vertically (the Y axis).

Enter the table at ±.0035 at the right side. Move along the horizontal grid to the bold 45° line. Here the horizontal line intersects a particular arc. Follow this arc counterclockwise to the scale on the left side. Read the equivalent positional tolerance: ⌀.010.

B.3 Conversion of Square Tolerance Zone to Positional Tolerance

The square and the circumscribed circle in the lower left corner of Table B-1 show graphically the geometric relationship between a square tolerance zone and the equivalent circular positional tolerance zone. To use the table to convert a square tolerance zone to positional tolerance, first translate the given square tolerance zone to a bilateral tolerance (divide by two); then read the table as in Section B.2.

Example: A square tolerance of .007 is the equivalent of a bilateral tolerance of ±.0035, which converts to a positional tolerance of ⌀.010.

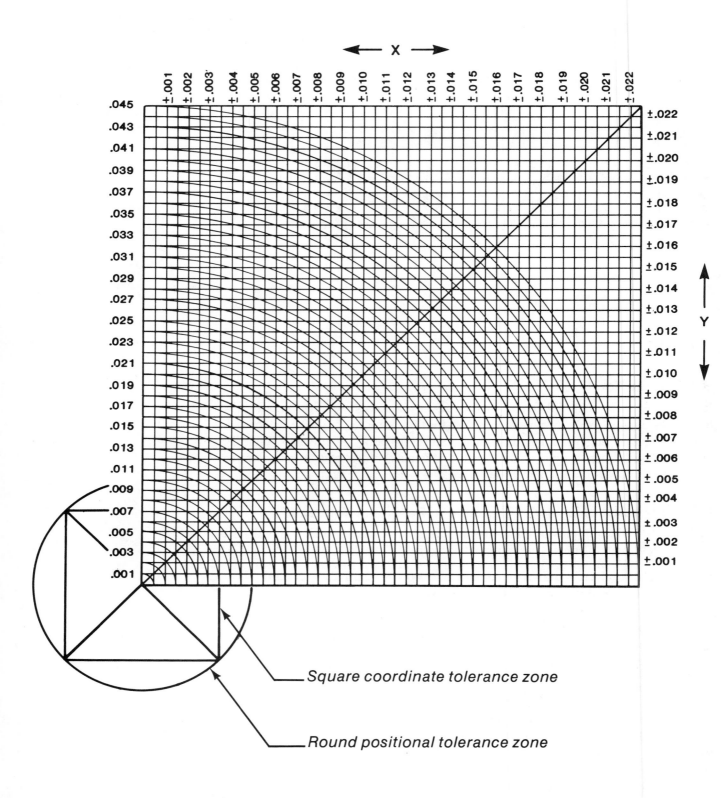

Table B-1   Conversion of bilateral tolerance to positional tolerance.

B.4 Conversion of Rectangular Tolerance Zone to Positional Tolerance

Table B-1 also can be used to convert a rectangular tolerance zone to a positional tolerance zone. The drawing below shows graphically the geometric relationship between the two kinds of tolerance zones.

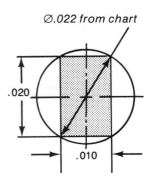

To use the table to convert a rectangular tolerance zone to positional tolerance, first translate the given rectangular tolerance zone to bilateral tolerances.

Example: Convert a .010 × .020 rectangular tolerance zone to the equivalent diametral positional tolerance zone.

Enter the table at the top at ±.005 (the bilateral tolerance equivalent of a .010-wide tolerance zone) and at the right side at ±.010 (the bilateral tolerance equivalent of a .020-high tolerance zone). Where the vertical ±.005 line and the horizontal ±.010 line intersect, find the closest arc. Follow that arc counterclockwise to the scale on the left side and read ∅.022. The true value is somewhere between .022 and .023 (actually it is .0224), but three-place accuracy is adequate.

B.5 Conversion of Coordinate Measurements to Positional Tolerance

The drawing below Table B-2 shows a typical situation for one feature located by basic dimensions from two datums. The location of the feature is inspected by a coordinate measurement machine from the two datums on the actual part, and the difference between the actual measurement and the basic dimension in each direction is noted.

In the drawing, the differences are labelled X difference and Y difference. The Z diameter is the corresponding positional tolerance diameter. All the figures in the body of the table are Z values.

Example: The drawing of a plate requires that a hole of diameter .374–.380 must be in true position relative to two datums within ∅.014 at MMC.

The location of the hole is measured, and the axis is found to be .005 from true position in the X direction and .006 off in the Y direction. Should the part be accepted or rejected?

Enter the table at the .005 column on the X scale and move up to the row headed .006 on the Y scale. Find the Z value, .0156. This is the diameter of the equivalent positional tolerance zone. If the tolerance were applied RFS, the part would be rejected. However, since the tolerance is specified MMC, a bonus tolerance may make the part acceptable if the hole as produced is larger than its MMC.

POSITIONAL TOLERANCE DIAMETER Z																				
Y \ X	.001	.002	.003	.004	.005	.006	.007	.008	.009	.010	.011	.012	.013	.014	.015	.016	.017	.018	.019	.020
.020	.0400	.0402	.0404	.0408	.0412	.0418	.0424	.0431	.0439	.0447	.0456	.0466	.0477	.0488	.0500	.0512	.0525	.0538	.0552	.0566
.019	.0380	.0382	.0385	.0388	.0393	.0398	.0405	.0412	.0420	.0429	.0439	.0449	.0460	.0472	.0484	.0497	.0510	.0523	.0537	.0552
.018	.0360	.0362	.0365	.0369	.0374	.0379	.0386	.0394	.0403	.0412	.0422	.0433	.0444	.0456	.0469	.0482	.0495	.0509	.0523	.0538
.017	.0340	.0342	.0345	.0349	.0354	.0360	.0368	.0376	.0385	.0394	.0405	.0416	.0428	.0440	.0453	.0467	.0481	.0495	.0510	.0525
.016	.0321	.0322	.0325	.0330	.0335	.0342	.0349	.0358	.0367	.0377	.0388	.0400	.0412	.0425	.0439	.0452	.0467	.0482	.0497	.0512
.015	.0301	.0303	.0306	.0310	.0316	.0323	.0331	.0340	.0350	.0360	.0372	.0384	.0397	.0410	.0424	.0439	.0453	.0469	.0484	.0500
.014	.0281	.0283	.0286	.0291	.0297	.0305	.0313	.0322	.0333	.0344	.0356	.0369	.0382	.0396	.0410	.0425	.0440	.0456	.0472	.0488
.013	.0261	.0263	.0267	.0272	.0278	.0286	.0295	.0305	.0316	.0328	.0340	.0354	.0368	.0382	.0397	.0412	.0428	.0444	.0460	.0477
.012	.0241	.0243	.0247	.0253	.0260	.0268	.0278	.0288	.0300	.0312	.0325	.0339	.0354	.0369	.0384	.0400	.0416	.0433	.0449	.0466
.011	.0221	.0224	.0228	.0234	.0242	.0250	.0261	.0272	.0284	.0297	.0311	.0325	.0340	.0356	.0372	.0388	.0405	.0422	.0439	.0456
.010	.0201	.0204	.0209	.0215	.0224	.0233	.0244	.0256	.0269	.0283	.0297	.0312	.0328	.0344	.0360	.0377	.0394	.0412	.0429	.0447
.009	.0181	.0184	.0190	.0197	.0206	.0216	.0228	.0241	.0254	.0269	.0284	.0300	.0316	.0333	.0350	.0367	.0385	.0402	.0420	.0439
.008	.0161	.0165	.0171	.0179	.0189	.0200	.0213	.0226	.0241	.0256	.0272	.0288	.0305	.0322	.0340	.0358	.0376	.0394	.0412	.0431
.007	.0141	.0146	.0152	.0161	.0172	.0184	.0198	.0213	.0228	.0244	.0261	.0278	.0295	.0313	.0331	.0349	.0368	.0386	.0405	.0424
.006	.0122	.0126	.0134	.0144	.0156	.0170	.0184	.0200	.0216	.0233	.0250	.0268	.0286	.0305	.0323	.0342	.0360	.0379	.0398	.0418
.005	.0102	.0108	.0117	.0128	.0141	.0156	.0172	.0189	.0206	.0224	.0242	.0260	.0278	.0297	.0316	.0335	.0354	.0374	.0393	.0412
.004	.0082	.0089	.0100	.0113	.0128	.0144	.0161	.0179	.0197	.0215	.0234	.0253	.0272	.0291	.0310	.0330	.0349	.0369	.0388	.0408
.003	.0063	.0072	.0085	.0100	.0117	.0134	.0152	.0171	.0190	.0209	.0228	.0247	.0267	.0286	.0306	.0325	.0345	.0365	.0385	.0404
.002	.0045	.0056	.0072	.0089	.0108	.0126	.0146	.0165	.0184	.0204	.0224	.0243	.0263	.0283	.0303	.0322	.0342	.0362	.0382	.0402
.001	.0028	.0045	.0063	.0082	.0102	.0122	.0141	.0161	.0181	.0201	.0221	.0241	.0261	.0281	.0301	.0321	.0340	.0360	.0380	.0400

X →

Table B-2   Conversion of coordinate measurements to positional tolerance.

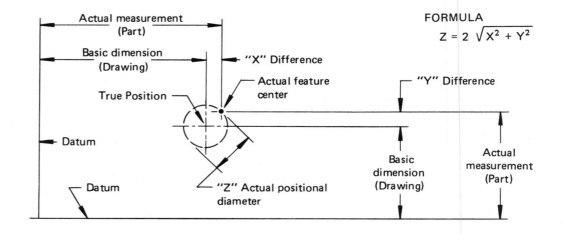

FORMULA

$$Z = 2 \sqrt{X^2 + Y^2}$$

The actual diameter of the hole is measured and is found to be .377 diameter. This exceeds the MMC (.374) by .003, the bonus tolerance. Add .003 to the .014-diameter positional tolerance given on the print and obtain .017 diameter as the *total* permissible positional tolerance. Since the actual hole location is within a .0156-diameter circle, it is inside the .017-diameter limit, and the part is acceptable.

The conversion method described above works equally well for any number of features located by positional tolerances. Table B-2 can be used to verify the location of feature patterns such as hole patterns that are specified on drawings with composite positional tolerances (Section 18.11). In this case, two conversions must be made: one for the location of the pattern as a whole and the other for the location of the individual holes within the pattern.

INDEX

CHAPTER REVIEW PROBLEMS/
COMPREHENSIVE EXERCISES

CHAPTER 1 REVIEW PROBLEMS

Write the letter of the best answer in the space at left.

Geometric tolerance is the ___(1)___ variation in the ___(2)___ of a feature of an object and its ___(3)___ other features.

_____ 1. (a) partial   (b) total   (c) bilateral   (d) unilateral

_____ 2. (a) shape   (b) size   (c) limit   (d) direction

_____ 3. (a) boundary with   (b) proximity to
               (c) normality to   (d) relationship to

_____ 4. When geometric tolerances are not specified on a drawing, the geometry of the object is controlled by the _____.

               (a) dimensions   (b) views   (c) material   (d) scale

_____ 5. The method of specifying geometric tolerances recommended by the Y14.5M standard is by the use of _____.

               (a) local notes   (b) general notes
               (c) symbols   (d) size tolerances

_____ 6. The shape of an object and the relationship of its features is called its _____.

               (a) tolerance   (b) orientation   (c) physics   (d) geometry

_____ 7. If the upper surface of a block is located by a ±.010 tolerance from a bottom surface that is absolutely flat, the upper surface can be curved by a maximum of _____.

               (a) .005   (b) .010   (c) .020   (d) more than .020

_____ 8. Geometric tolerance symbols were originally developed by the British during the late _____.

               (a) 1920s   (b) 1930s   (c) 1940s   (d) 1950s

_____ 9. The name of the organization which has standardized American geometric tolerance practices is abbreviated _____.

               (a) NASA   (b) ASA   (c) ANSI   (d) ASI

_____ 10. The latest revision of the national standard on dimensioning and tolerancing was made in _____.

               (a) 1966   (b) 1973   (c) 1980   (d) 1982

CHAPTER 2 REVIEW PROBLEMS

Next to each geometric characteristic listed below, sketch the appropriate symbol, selecting from the symbols shown.

_____ 1. Straightness   _____ 8. Flatness

_____ 2. Roundness   _____ 9. Cylindricity

_____ 3. Profile of a line   _____ 10. Profile of a surface

_____ 4. Parallelism   _____ 11. Perpendicularity

_____ 5. Angularity   _____ 12. Circular runout

_____ 6. Concentricity   _____ 13. Total runout

_____ 7. Positional tolerance

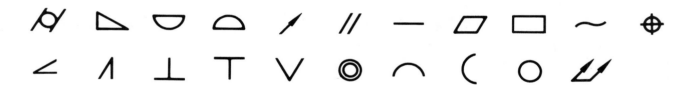

_____ 14. The symbol for *diameter* is one of those shown below. Sketch the correct symbol in the space at left.

ϕ $\ominus$ $\varnothing$ D

Write the letter of the best answer in the space at left.

_____ 15. Which of the following types of lines cannot have a feature control frame attached to it?

     (a) extension   (b) leader   (c) center   (d) dimension

_____ 16. A datum feature symbol is drawn _____.

     (a) .25 × 1.25   (b) .31 × .62
     (c) .38 × .62   (d) .31 × 1.50

_____ 17. A feature control frame is drawn with _____.

     (a) thin lines   (b) thick lines
     (c) phantom lines   (d) broken lines

(over)

_____ 18. A feature control frame is drawn _____.

(a) .25 × .75   (b) .31 × .62

(c) .31 × 1.25 min   (d) .31 × any length

_____ 19. Modifier symbols that are drawn in a circle have a circle diameter of _____.

(a) .12   (b) .16   (c) .24   (d) .31

_____ 20. The various parts of a feature control frame are separated by _____.

(a) vertical lines   (b) diagonal lines

(c) dashes   (d) hyphens

CHAPTER 3 REVIEW PROBLEMS   PART I

Write the letter of the best answer in the space at left.

_____ 1. The area taken up by the total permissible error in a geometric form is the _____.

(a) virtual condition   (b) basic dimension
(c) datum   (d) tolerance zone

_____ 2. The maximum material condition of a hole is always the _____ size.

(a) smallest   (b) largest   (c) nominal   (d) basic

_____ 3. The maximum material condition of a shaft is always the _____ size.

(a) smallest   (b) largest   (c) nominal   (d) basic

_____ 4. The symbol for *regardless of feature size* is _____.

(a) Ⓡ   (b) Ⓕ   (c) Ⓢ   (d) Ⓡ̲Ⓕ̲Ⓢ̲

_____ 5. The virtual condition of an object is the size that results in the _____ fit with a mating part.

(a) loosest   (b) average   (c) tightest   (d) nominal

_____ 6. The total reading of the pointer of a dial indicator when used to test a geometric tolerance is abbreviated _____.

(a) FIR   (b) FIM   (c) TIR   (d) TIM

_____ 7. The symbol for a geometric tolerance zone that is transferred from the thickness of the part being controlled to the thickness of the mating part is _____.

(a) Ⓟ   (b) Ⓜ   (c) Ⓢ   (d) Ⓔ

_____ 8. A theoretically exact surface from which a dimension may be taken is a _____.

(a) feature   (b) datum
(c) basic surface (d) nominal surface

_____ 9. Any line, real or imaginary, that can be drawn on a surface is called a (an) _____.

(a) element   (b) feature   (c) datum   (d) radial line

_____ 10. A line that extends toward or away from a center is called a (an) _____.

(a) element   (b) center line
(c) extension line (d) radial line

CHAPTER 3 REVIEW PROBLEMS   PART II

In the space to the right of each statement, neatly letter the word or term from the list below which pertains to the statement.

Any line, real or imaginary, that may be drawn on a surface.

1. _____

Any portion of an object, such as a point, axis, plane, or cylindrical surface, or a tab, recess, or groove.

2. _____

Describes a dimension that does not have a tolerance. The tolerance is given elsewhere.

3. _____

The size of a shaft as produced is such that it contains the most steel possible.

4. _____

A theoretically exact surface or line from which dimensions or geometric tolerances may be taken.

5. _____

The geometric tolerance specified applies no matter how big or small the feature may be produced.

6. _____

The size of an object that results in the tightest fit with a mating part.

7. _____

The total reading of a dial indicator.

8. _____

Pointing toward or away from a center.

9. _____

The area taken up by the total amount of permissible geometric error.

10. _____

Select answers from the following:

TOLERANCE ZONE	ELEMENT	NORMALITY
VIRTUAL CONDITION	RADIAL LINE	MMC
DATUM	FEATURE	FIM
BASIC	CHARACTERISTIC	RFS
MEDIAN	MODIFIER	LMC

CHAPTER 4 REVIEW PROBLEMS

Write the letter of the best answer in the space at left.

A datum is a theoretically exact ___(1)___ from which ___(2)___ may be taken, or the geometric form or ___(3)___ of another feature may be determined.

_____ 1. (a) feature   (b) element   (c) shape   (d) intersection

_____ 2. (a) tolerances   (b) dimensions   (c) views   (d) squareness

_____ 3. (a) length   (b) width   (c) size   (d) position

_____ 4. The terms *datum plane* and *datum feature* are _____.

   (a) interchangeable but not identical
   (b) identical in meaning
   (c) ambiguous
   (d) not identical in meaning

_____ 5. A good feature to select for a datum would be a (an) _____.

   (a) lathe center   (b) bearing surface
   (c) center line   (d) axis

_____ 6. If more than one datum is used, the one that determines the fit of the part is called the _____.

   (a) first datum   (b) leading datum
   (c) auxiliary datum   (d) basic datum

_____ 7. The datum feature symbol is drawn _____.

   (a) .25 × .50   (b) .31 × .62   (c) .38 × .75   (d) .50 × 1.00

_____ 8. The datum feature symbol is drawn with _____.

   (a) phantom lines   (b) broken lines
   (c) thin lines   (d) thick lines

_____ 9. Certain letters are not used in the datum feature symbols. These letters are _____.

   (a) I, O, Q   (b) A, B, C   (c) M, S, P   (d) X, Y, Z

_____ 10. The datum feature is a feature of the _____.

   (a) part shown on the drawing
   (b) the tooling used to manufacture the part
   (c) neither of the above
   (d) both of the above

(over)

_____ 11. The datum plane exists only in the _____.

(a) drawing
(b) part shown on the drawing
(c) tooling used to manufacture the part
(d) ANSI standard

_____ 12. When an object rests on a surface plate, contact is made only at three _____.

(a) high points (b) corners (c) edges (d) defects

CHAPTER 5 REVIEW PROBLEMS

Write the letter of the best answer in the space at left.

_____ 1. A dial indicator is a device used to detect movements as small as _____.

(a) 1 cm   (b) 1 mm   (c) .001 in.   (d) .0001 in.

_____ 2. A dial indicator is similar in appearance to a small pocket watch, except that it has _____.

(a) no bezel   (b) no "hands"   (c) one "hand"   (d) no stem

_____ 3. In making a dial indicator test for parallelism, the FIM is .004. This indicates a parallelism error of _____.

(a) .002   (b) .004   (c) .008   (d) more than .008

_____ 4. The part of a dial indicator that is rotatable is called the _____.

(a) clamp   (b) case   (c) probe   (d) bezel

_____ 5. If the movement observed on a dial indicator is .001 to the left and .003 to the right, the FIM is _____.

(a) .001   (b) .003   (c) .004   (d) .008

_____ 6. The "F" in FIM stands for _____.

(a) free   (b) fine   (c) final   (d) full

_____ 7. A dial indicator cannot be used for testing _____.

(a) concentricity   (b) straightness
(c) runout   (d) surface finish

_____ 8. A dial indicator would most likely be used for testing parts made in lots of _____.

(a) 10   (b) 1000   (c) 5000   (d) 10,000

_____ 9. A special gage used for inspecting a machined part simulates the fit of the _____.

(a) mating part   (b) dial indicator
(c) surface plate   (d) vee-block

_____ 10. The pointer of a dial indicator moves _____.

(a) clockwise only
(b) counterclockwise only
(c) clockwise and counterclockwise
(d) radially

CHAPTER 6 REVIEW PROBLEMS   PART I

Complete the five general rules governing geometric tolerancing, choosing words or terms from the list below. Place the letter of each correct word or term in the space at left. The same answer may be used more than once.

_____   1.   When no geometric tolerance is specified, the _____ tolerance controls the form of the part as well as the size.

_____   2.   For _____ tolerance, a modifier must always be specified.

_____   3.   When no modifier is specified, all geometric characteristics except _____ tolerance apply RFS.

_____   4.   All geometric tolerances specified for a screw thread apply to the _____ diameter.

_____   5.   A size feature that is used as a datum and is controlled by its own geometric tolerance applies at its ___(5)___ even

_____   6.   though ___(6)___ is specified.

(a) concentricity   (h) modifier
(b) dimensional   (i) outside
(c) external   (j) parallelism
(d) internal   (k) pitch
(e) LMC   (l) positional
(f) major   (m) runout
(g) MMC   (n) virtual condition

_____   7.   Rule 1 also states that no element of a part shall extend beyond the boundary of the _____.

(a) virtual condition   (b) MMC
(c) LMC   (d) maximum size

_____   8.   Rule 1 applies only to _____.

(a) individual features   (b) interrelated features
(c) datums   (d) stock items

_____   9.   In addition to screw threads, Rule 4 also applies to splines and

_____.

(a) gears   (b) pulleys   (c) knurling   (d) shafts

_____   10. Where it is not intended for Rule 5 to apply to a particular datum, the designer specifies a zero tolerance _____.

(a) at LMC   (b) at MMC   (c) ⓟ   (d) RFS

CHAPTER 6 REVIEW PROBLEMS   PART II

In the space below, draw a bar, full size, that is 49.05—50.05 mm wide (left to right) and 12.00—12.05 mm high. Add limit dimensions.

On a bar produced from this drawing, what will be the sizes of the height and width at the MMC boundary?

1. _____ × _____

When the part drawn above is produced at its MMC boundary, what is the permissible error in its geometric form (straightness, flatness, parallelism, and perpendicularity)?

2. _____

The positional tolerance shown below is incomplete. What is lacking?

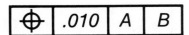

3. _____

In the straightness tolerance shown below, no modifier is specified. Which modifier automatically applies?

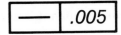

4. _____

In the space below, draw the feature control frame for a screw thread whose pitch diameter is to be concentric with Datum A within .002 inch. No screw thread drawing is necessary.

5.

In the space below, draw the feature control frame for a gear whose pitch diameter is to be concentric with Datum A within .001 inch. No gear drawing is necessary.

6.

In the drawing below, the equally spaced holes are specified in positional tolerance relative to Datum A within ∅ .010 inch when the equally spaced holes *and* Datum A are at MMC. However, Rule 5 states that in this situation (Datum A is a *size* feature with its own positional tolerance) the datum applies at its *virtual* condition even though MMC is specified. What is the virtual size of Datum A?

7. _____

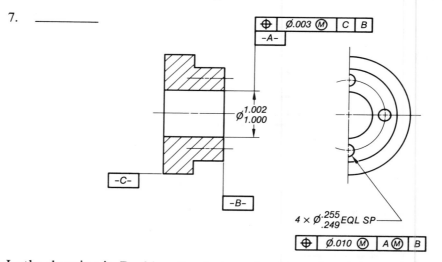

In the drawing in Problem 7, what can be done when it is not intended for the virtual condition to apply?

(a) Increase the positional tolerance for the datum.

(b) Decrease the positional tolerance for the datum.

(c) Omit the positional tolerance for the datum.

(d) Specify zero tolerance at MMC for the datum.

8. _____

CHAPTER 7 REVIEW PROBLEMS   PART I

Write the letter of the best answer in the space at left.

_____ 1. Straightness error is the measure of how much each _____ in a surface deviates from being a straight line.

(a) feature (b) element (c) chord (d) datum

_____ 2. There are two varieties of straightness: element straightness and _____ straightness.

(a) size-feature (b) cylindrical (c) square (d) conical

_____ 3. The shape of the tolerance zone for element straightness is _____.

(a) a cylinder (b) a cone (c) a ring
(d) the space between two parallel straight lines

_____ 4. Unless otherwise specified, all straightness tolerances apply _____.

(a) at MMC (b) at LMC (c) RFS (d) to a limited length

_____ 5. In size-feature straightness, the straightness tolerance and the size tolerance are interrelated. Therefore the boundary of the size feature _____ extend beyond the MMC.

(a) can (b) cannot (c) should (d) should not

_____ 6. A tolerance that specifies straightness per a certain distance is called _____ straightness.

(a) digital (b) unit (c) dimensional (d) partial

_____ 7. If the same straightness tolerance is applied to a surface in two directions (in two views of the drawing), this is the same as a _____.

(a) flatness tolerance (b) dimensional tolerance
(c) total tolerance (d) datum plane

_____ 8. In size-feature straightness for a cylinder or cone, the tolerance zone is a _____.

(a) diameter (b) rectangle (c) cone (d) ring

_____ 9. The feature control frame for straightness of a flat surface is applied in a view that shows the surface _____.

(a) in the front view (b) in a side view
(c) as an area (d) as a straight line

_____ 10. When specifying unit straightness, care must be taken that the _____ straightness error is not excessive.

(a) partial (b) total
(c) perpendicular (d) opposite surface

CHAPTER 7 REVIEW PROBLEMS   PART II

In the space below, draw the feature control frame for a straightness tolerance of .002 inch.

1.

In the space below, draw two views of a part that is 2 inches wide (left to right), .75 inch high, and .50 inch deep. Omit dimensions. Add a .005-inch element straightness tolerance to the top surface in the front view and a .002-inch element straightness tolerance to the rear surface.

2.

In the space below, repeat the front view of the drawing in Problem 2 and add a two-place height dimension. Show with a properly placed feature control frame that the height dimension is to be straight within .003 inch.

3.

(over)

A long round shaft has a unit size straightness of .001 inch per 1.000 inch length regardless of feature size. Draw the appropriate feature control frame below. Hint: the tolerance zone for the size of a cylinder is a special shape.

4.

A long round shaft has a unit size straightness of .001 inch per 1.000 inch length at MMC, but the total straightness at MMC must be within .010 inch. Draw the appropriate feature control frame below.

5.

CHAPTER 8 REVIEW PROBLEMS   PART I

Write the letter of the best answer in the space at left.

_____ 1. A flatness tolerance zone is the space between two parallel _____.

(a) planes   (b) elements   (c) straight lines   (d) datums

_____ 2. Flatness is specified by directing the feature control frame to the view in which the surface is _____.

(a) true size   (b) horizontal   (c) vertical   (d) an edge view

_____ 3. The measure of how much a plane surface deviates from being a true plane is called _____.

(a) plane deviation   (b) plane variation
(c) flatness error   (d) a tolerance zone

_____ 4. A surface can be straight in one direction but not be flat. The preceding statement _____.

(a) requires more data   (b) is questionable
(c) is true   (d) is false

_____ 5. No element of a surface controlled by a flatness tolerance may extend beyond the _____ boundary of the part.

(a) LMC   (b) MMC   (c) RFS   (d) projected

_____ 6. A flatness tolerance which specifies tolerance per a certain distance is called _____.

(a) dimensional flatness   (b) unit flatness
(c) partial flatness   (d) digital flatness

_____ 7. In specifying flatness only in a particular area, the area is section-lined on the drawing and surrounded by _____.

(a) a break line   (b) extension lines
(c) hidden lines   (d) chain lines

_____ 8. Flatness in a specified area may be used on a _____.

(a) large casting   (b) cylindrical part
(c) precision machined part   (d) small sheet metal part

CHAPTER 8 REVIEW PROBLEMS   PART II

In the space below, draw the feature control frame for a flatness tolerance of .002 inch.

1.

In the space below, draw two views of a part that is 2 inches wide (left to right), .75 inch high, and .50 inch deep. Omit dimensions. Add a flatness tolerance of .005 inch to the top surface and a flatness tolerance of .002 inch to the rear surface.

2.

In the drawing below, with no flatness control specified, what is the maximum permissible flatness error of the top surface if the bottom surface is perfectly flat?

3. _____

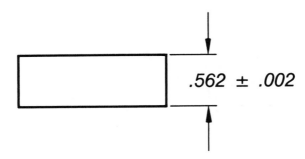

.562 ± .002

In the drawing for Problem 3, what is the maximum permissible flatness error of the top surface if the bottom surface is out of flat by .001 inch?

4. _____

In the space below, redraw the drawing for Problem 3 and add flatness tolerances of .001 inch to the top and bottom surfaces.

5.

Below is shown, enlarged, one possible interpretation of the student's drawing for Problem 5. Given the size and flatness tolerances of the correct drawing and this interpretation, fill in the values of dimensions A through E in the spaces provided.

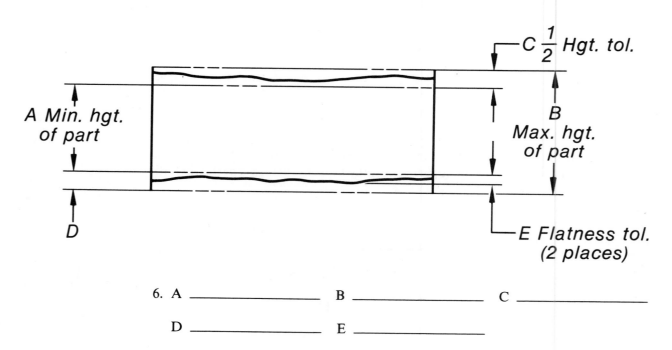

6. A _____   B _____   C _____

   D _____   E _____

CHAPTER 9 REVIEW PROBLEMS   PART I

Write the letter of the best answer in the space at left.

_____ 1. A roundness tolerance controls only circular _____.

(a) features   (b) surfaces   (c) elements   (d) datums

_____ 2. A roundness tolerance _____ be within the dimensional
size limits of the part (Rule 1).

(a) may   (b) may not   (c) must   (d) must not

_____ 3. A roundness tolerance always applies _____.

(a) FIM (Rule 1)   (b) at LMC (Rule 2)
(c) RFS (Rule 3)   (d) at MMC (Rule 4)

_____ 4. The shape of the tolerance zone for roundness is most like
_____.

(a) a cylinder   (b) a radius
(c) an ellipse   (d) a ring or washer

_____ 5. Roundness error is measured as the _____ space between
two concentric perfect circles.

(a) radial   (b) diametral   (c) angular   (d) circular

_____ 6. A roundness tolerance can be thought of as a _____ tolerance curled into a circle.

(a) concentricity   (b) cylindricity
(c) straightness   (d) flatness

_____ 7. The only accurate method of testing for roundness is by placing
the part _____.

(a) on a turntable   (b) on lathe centers
(c) in a vee-block   (d) on a surface plate

CHAPTER 9 REVIEW PROBLEMS   PART II

All the dimensions in this exercise are in metric units (millimeters).

In the space below, draw the feature control frame for a roundness tolerance of .05 mm.

1.

In the space below, draw two views of a cylinder 15 mm in diameter and 10 mm long. Space the views about 40 mm (1½ inches) apart. Omit dimensions. Add a roundness tolerance of .05 mm.

2.

In the drawing below, with no roundness control specified, what is the maximum permissible roundness error of the cylindrical surface?

3. _____

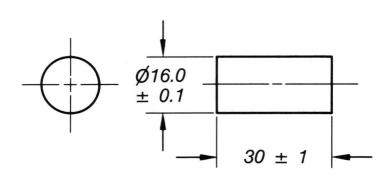

In the space below, redraw the drawing for Problem 3 and add a roundness tolerance of .02 mm to the cylindrical surface.

4.

(over)

Below is shown, enlarged, one possible interpretation of the student's drawing for Problem 4. Given the size and roundness tolerance of the correct drawing and this interpretation, fill in the values of dimensions A through D in the spaces provided.

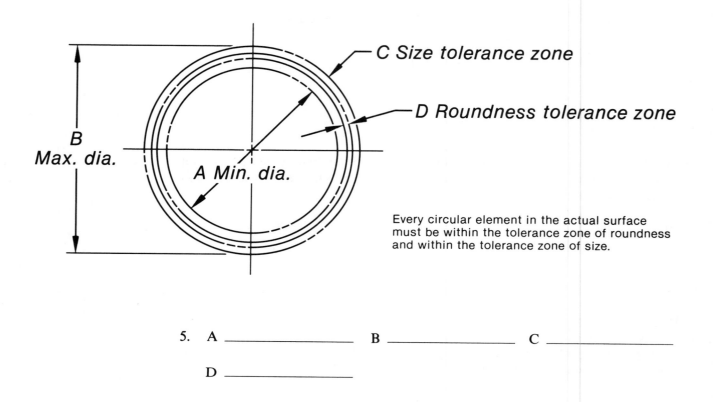

C Size tolerance zone

D Roundness tolerance zone

B Max. dia.

A Min. dia.

Every circular element in the actual surface must be within the tolerance zone of roundness and within the tolerance zone of size.

5. A _____ B _____ C _____

D _____

CHAPTER 10 REVIEW PROBLEMS   PART I

Write the letter of the best answer in the space at left.

_____   1.   Cylindricity is a combination of roundness and _____.
                   (a) flatness   (b) straightness
                   (c) parallelism   (d) perpendicularity

_____   2.   The shape of a cylindrical tolerance zone is the _____ space
                   between two concentric perfect cylinders.
                   (a) radial   (b) diametral   (c) total   (d) angular

_____   3.   The shape of the tolerance zone for cylindricity is most like a
                   _____.
                   (a) doughnut   (b) globe   (c) tube   (d) cylinder

_____   4.   A cylindrical tolerance _____ be within the dimensional
                   size limits of the part (Rule 1).
                   (a) may   (b) may not   (c) must   (d) must not

_____   5.   A cylindricity tolerance always applies _____.
                   (a) FIM (Rule 1)   (b) at LMC (Rule 2)
                   (c) RFS (Rule 3)   (d) at MMC (Rule 4)

_____   6.   A cylindricity tolerance can be thought of as a _____
                   tolerance curled into the shape of a cylinder.
                   (a) straightness   (b) flatness   (c) concentricity   (d) profile

_____   7.   Cylindricity is inspected by the same methods used for round-
                   ness, except that the longitudinal elements must be tested for
                   _____.
                   (a) concentricity   (b) straightness
                   (c) flatness   (d) coaxiality

CHAPTER 10 REVIEW PROBLEMS   PART II

In the space below, draw the feature control frame for a cylindricity tolerance of .0005 inch.

1.

In the space below, draw two views of a cylinder ⅝ inch in diameter and 1 inch long. Space the views about 1½ inches apart. Omit dimensions. Add a cylindricity tolerance of .003 inch.

2.

In the drawing below, with no cylindricity control specified, what is the maximum permissible cylindricity error?

3.
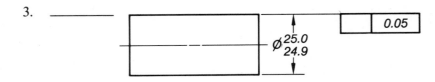

In the space below, redraw the drawing for Problem 3 and add a cylindrical tolerance of .002 inch.

4.

No modifer symbol should be specified in any of the geometric tolerances used in the chapter review problems thus far. However, a particular modifier applies to *all* of them, although it is never specified in the feature control frames. What is that modifier?

5.   The whole name: _____

      The abbreviation: _____

      The symbol: _____

CHAPTER 11 REVIEW PROBLEMS   PART I

Write the letter of the best answer in the space at left.

_____   1. Profile of a line tolerance controls not an entire surface but
                     only _____.
                     (a) chords (b) radii (c) elements (d) arcs

_____   2. Unless otherwise specified, a profile tolerance is assumed to be
                     _____.
                     (a) unilateral (b) bilateral (c) unidirectional (d) aligned

_____   3. A dimension showing the profile tolerance on the drawing,
                     exaggerated in size, is necessary when the tolerance is specified
                     as _____.
                     (a) unilateral (b) bilateral (c) unidirectional (d) aligned

_____   4. Profile of a line and profile of a surface tolerance are
                     _____.
                     (a) related (b) identical (c) unrelated (d) redundant

_____   5. Profile of a surface tolerance controls the profile in _____.
                     (a) one direction   (b) two directions
                     (c) three directions   (d) one curved line

                 6. Make a neat sketch in the space below showing a profile of a
                     surface tolerance. Add a leader and indicate that the tolerance
                     applies "all around," not just to a limited area.

_____   7. Profile of a surface tolerance is most similar to _____, ex-
                     cept that it applies to a curved surface.
                     (a) straightness (b) flatness
                     (c) roundness   (d) parallelism

_____   8. Profile of a line tolerance is most similar to _____, except
                     that it applies to curved elements.
                     (a) straightness (b) flatness
                     (c) cylindricity (d) parallelism

_____   9. When a profile tolerance for a particular feature requires that
                     it be related to another feature, the drafter specifies a
                     _____.
                     (a) unilateral tolerance   (b) bilateral tolerance
                     (c) positional tolerance   (d) datum

(over)

_____ 10. When profile of a surface tolerance is used to control coplanarity, a leader directs the feature control symbol to _____.

(a) an extension line between the surfaces
(b) the larger surface
(c) a dimension line locating the surfaces
(d) the surface that is to be machined first

CHAPTER 11 REVIEW PROBLEMS   PART II

On the drawing below, show that the *curved elements* are within .1 mm, total bilateral tolerance, of the desired profile. The profile tolerance applies between points X and Y.

1.

Redraw below the figure in Problem 1, to make the tolerance apply "all around."

2.

Write local notes to replace the symbology in Problem 1 and Problem 2.

3. Problem 1 _____

   _____

   _____

4. Problem 2 _____

   _____

   _____

(over)

On the drawing below, show that the entire curved surface is to have a profile tolerance of .1 mm only *inward* of the desired profile.

5.

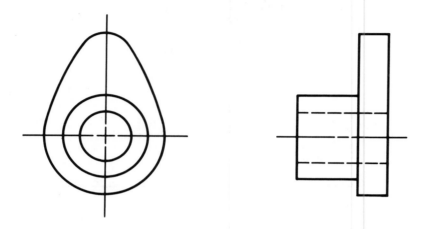

On the same cam, repeated below, show that the entire curved surface is to have a profile tolerance of .05 mm only *outward* of the desired profile.

6.

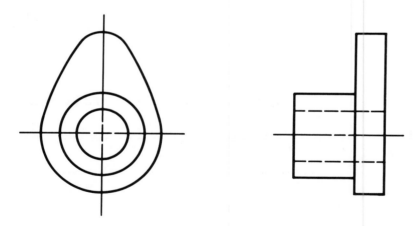

CHAPTER 12 REVIEW PROBLEMS   PART I

Write the letter of the best answer in the space at left.

_____   1. Which of the following geometric characteristics does not
                  require a datum?
                  (a) cylindricity   (b) parallelism
                  (c) perpendicularity   (d) concentricity

_____   2. Two surfaces are parallel if every point on each surface is
                  _____ the other surface.
                  (a) the same distance from   (b) beside
                  (c) outside   (d) close to

_____   3. Depending on the form of the controlled feature and the
                  datum, a parallelism tolerance zone may take either (any) of
                  _____.
                  (a) two forms   (b) three forms
                  (c) four forms   (d) five forms

_____   4. Any tolerance zone that is cylindrical in shape is designated in
                  the feature control frame by a symbol meaning _____.
                  (a) parallel (b) cylindricity (c) diameter (d) area

_____   5. When features of size are involved and no modifier is specified,
                  a parallelism tolerance applies _____.
                  (a) at MMC   (b) at LMC
                  (c) RFS   (d) at virtual condition

_____   6. The statement in Problem 5, above, is an application of
                  _____.
                  (a) Rule 1   (b) Rule 2   (c) Rule 3   (d) Rule 4

_____   7. The controlled feature in a parallelism tolerance must be
                  within the limits of the location and/or size tolerance. This is
                  an application of _____.
                  (a) Rule 1   (b) Rule 2   (c) Rule 3   (d) Rule 4

_____   8. If a surface is specified parallel to a datum within .002 inch and
                  no flatness is specified, the permissible flatness error of the
                  surface is _____.
                  (a) .0005 (b) .001 (c) .002 (d) .004

_____   9. When the MMC symbol is specified for a parallelism toler-
                  ance, the controlled feature is _____.
                  (a) parallel to a datum   (b) as large as possible
                  (c) as accurate as possible   (d) a size feature

(over)

_____ 10. The MMC symbol in a parallelism specification means that additional parallelism error is allowed when the feature is _____ .

(a) at its MMC (b) not at its MMC
(c) too large   (d) too small

CHAPTER 12 REVIEW PROBLEMS   •   PART II

A surface of a part is parallel with the opposite surface within .005. In the space below, draw the appropriate feature control frame.

1.

A hole in a part is parallel with the flat base of the part within .002, regardless of the produced size of the hole. In the space below, draw the appropriate feature control frame.

2.

In the drawing below, the upper hole at **MMC** should be parallel with the lower hole within .002 diameter, regardless of the produced size of the lower hole. Add the appropriate symbology to the drawing.

3.

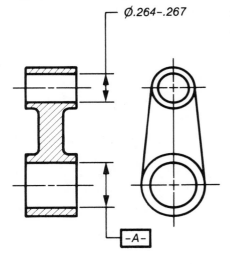

Ø.264–.267

–A–

In Problem 3, the parallelism tolerance will vary with the produced size of the upper hole. Complete the table below, showing this variation.

4.

Feature Size	Diameter Tolerance Zone Allowed
.264	.002
.265	_____
.266	_____
.267	_____

Under what conditions is a parallelism tolerance specified as a diameter? (Hint: What shape must the controlled feature be? What shape must the datum be?)

5. _____

_____

_____

In the space below, draw one view of a part with two parallel holes. Omit dimensions. Show that the hole on the right, regardless of feature size, is parallel with the other hole within .003. (Hint: What is the shape of the tolerance zone?)

6.

In problem 6, say that both holes were .375 +.005 –.000 diameter and the parallelism tolerance applies at the MMC of both holes. Parallelism is still within .003. Draw the appropriate feature control frame in the space below.

7.

For the drawing in Problem 6 with the conditions given in Problem 7, make up a tolerance variation table for six possible combinations of sizes, using the headings below.

8. Feature Size	Datum Size	Diametral Tolerance
_____	_____	_____
_____	_____	_____
_____	_____	_____
_____	_____	_____
_____	_____	_____
_____	_____	_____

CHAPTER 13 REVIEW PROBLEMS   PART I

Write the letter of the best answer in the space at left.

_____ 1. Two plane surfaces or straight lines are perpendicular when they are _____ each other.

(a) opposite (b) beside (c) 90° to (d) 45° to

_____ 2. The most common form of the tolerance zone for perpendicularity is the space between two imaginary _____, perfectly perpendicular to the datum.

(a) parallel planes (b) perpendicular planes
(c) flat surfaces (d) straight lines

_____ 3. A perpendicularity tolerance zone may be in the form of an imaginary cylinder, perfectly _____ the datum.

(a) concentric with (b) aligned with
(c) perpendicular to (d) parallel with

_____ 4. The diameter symbol (∅) is used with a perpendicularity tolerance when the form of the tolerance zone is a (an) _____.

(a) cylinder (b) circle (c) arc (d) axis

_____ 5. If no modifier is specified in the feature control frame, perpendicularity applies _____.

(a) at LMC (b) at MMC (c) at virtual condition (d) RFS

_____ 6. The statement in Problem 5 is an application of _____.

(a) Rule 1 (b) Rule 2 (c) Rule 3 (d) Rule 4

_____ 7. A perpendicularity tolerance may apply at MMC when the controlled feature is _____.

(a) a tab (b) a slot (c) a keyway (d) all of the above

_____ 8. A size feature with perpendicularity specified at MMC may exceed the dimensional limits because the size feature and its datum are _____ features.

(a) individual (b) interrelated (c) separate (d) adjacent

_____ 9. The statement in Problem 8 is an exception to _____.

(a) Rule 1 (b) Rule 2 (c) Rule 3 (d) Rule 4

_____ 10. A perpendicularity tolerance for a plane surface also controls _____ when it is not specified on the drawing.

(a) surface finish (b) roundness
(c) parallelism (d) flatness

CHAPTER 13 REVIEW PROBLEMS   PART II

A surface of a part is perpendicular to another surface within .003. In the space below, draw the appropriate feature control frame.

1.

A hole in a part is perpendicular with the flat base within .002, regardless of the produced size of the hole. In the space below, draw the appropriate feature control frame. (Hint: What is the shape of the tolerance zone?)

2.

In the part shown below, the hole at its MMC is perpendicular with the cylinder, RFS, within .004. Add the appropriate feature control frame.

3.

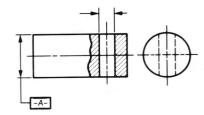

In Problem 3, say the hole is dimensioned .250–.255 diameter. The perpendicularity tolerance will vary with the produced size of the hole. Complete the tolerance variation table below for every one-thousandth variation of the hole size.

4.

Produced Size of Hole	Perpendicularity Tolerance
.250	∅.002
_____	_____
_____	_____
_____	_____
_____	_____
_____	_____

In Problem 4, what effect does the produced size of the cylinder have on the perpendicularity tolerance?

5. _____

(over)

Complete the following sentence, neatly lettering the correct words in the blank spaces.

6. A perpendicularity tolerance is specified as a diameter when the controlled feature is a _____ and the datum is a _____.

In the drawing below, show that the boss is perpendicular to the base within .002 at MMC. (Hint: What is the shape of the tolerance zone?)

7.

$\varnothing.500 \pm .002$

For the drawing in Problem 7, make up a tolerance variation table for five possible produced sizes of the boss. Use your own headings.

8.

The two drawings below show a part with a tab and a part with a mating slot.

9.

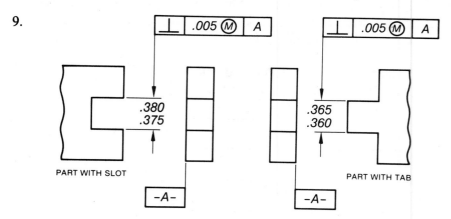

PART WITH SLOT   PART WITH TAB

(a) What is the MMC of the slot?   _____

(b) What is the MMC of the tab?   _____

(c) What is the virtual size of the slot?   _____

(d) What is the virtual size of the tab?   _____

(e) What is the minimum clearance between
    the tab and slot at the tightest fit?   _____

CHAPTER 14 REVIEW PROBLEMS   PART I

Write the letter of the best answer in the space at left.

_____ 1. A tolerance expressed in degrees results in a _____ tolerance zone.

(a) fan-shaped   (b) uniform   (c) radial   (d) conical

_____ 2. An angularity tolerance specifies the _____ width of a tolerance zone along the entire surface.

(a) angular   (b) fan-shaped   (c) uniform   (d) radial

_____ 3. The form of an angularity tolerance zone is the space between two _____ lines or planes at an exact angle to a datum.

(a) radial   (b) angular   (c) parallel   (d) basic

_____ 4. For angularity tolerance, a datum _____ be specified.

(a) may   (b) should   (c) should not   (d) must

_____ 5. The statement below is completed by the same word, used twice.

Angularity tolerance must be within the _____ size limits for flat surfaces and within _____ location limits for size features (holes, slots, and so forth).

(a) basic   (b) minimum   (c) maximum   (d) dimensional

_____ 6. The statement in Problem 5 is an application of _____.

(a) Rule 1   (b) Rule 2   (c) Rule 3   (d) Rule 4

_____ 7. When an angularity tolerance is specified, the angle dimension must be _____.

(a) basic   (b) unilateral   (c) bilateral   (d) unidirectional

_____ 8. Since angularity is a relationship characteristic, a _____ is always required.

(a) basic dimension   (b) datum
(c) tolerance   (d) limit dimension

_____ 9. When angularity is specified for symmetrical size features, the feature control frame and datum feature symbol _____ placed on the center line.

(a) are always   (b) are sometimes   (c) are never   (d) may be

_____ 10. Unless otherwise specified in the feature control frame, angularity applies _____.

(a) at LMC   (b) at MMC   (c) RFS   (d) at virtual condition

CHAPTER 14 REVIEW PROBLEMS   PART II

In the part shown below, the slanted surface is held at 30° to the base within .005. Complete the drawing.

1.

In the part shown below, the hole, regardless of feature size, is held at 60° to the base within .005. Complete the drawing.

2.

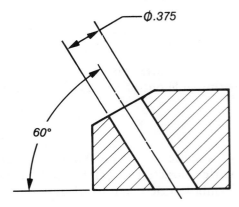

In the space below, draw the correct feature control frame for the drawing in Problem 2 if the angularity tolerance were held at the MMC of the hole.

3.

A .375–.380 diameter hole has an angularity tolerance of .006 at MMC relative to another hole, RFS. In the space below, draw a feature control frame to specify this situation.

4.

(over)

When the .375–.380 diameter hole in Problem 4 is *not* produced at its MMC, the angularity tolerance will not be .006. Make up a tolerance variation table below, giving the permissible angularity for each produced hole size, in thousandths.

5.   Produced Size   Angularity
           of Hole   Tolerance

        _____   _____

        _____   _____

        _____   _____

        _____   _____

        _____   _____

        _____   _____

In the drawing in Problem 1, the flatness of the slanted surface is not specified. What is the maximum permissible out-of-flatness of this surface?

6.   _____

Problems 4 and 5 deal with a size feature (a hole) that has an angularity tolerance at MMC. Another geometric characteristic can be used for this purpose in place of angularity. Draw the symbol of that characteristic.

7.   _____

CHAPTER 15 REVIEW PROBLEMS   PART I

Write the letter of the best answer in the space at left.

_____   1.   The word _____ is used to describe the general case of the geometric characteristic where axes are in line.

(a) concentricity   (b) alignment
(c) coaxiality   (d) runout

_____   2.   Three geometric characteristics that control the alignment of axes are concentricity, runout, and _____ tolerance.

(a) locational   (b) positional
(c) roundness   (d) circularity

_____   3.   Concentricity error is the amount by which the axes of two regular solids are _____.

(a) out of line   (b) in alignment   (c) curved   (d) irregular

_____   4.   Of the three geometric characteristics that are used to control coaxiality, the one used the least is _____.

(a) positional tolerance   (b) runout
(c) roundness   (d) concentricity

_____   5.   The form of the concentricity tolerance zone is an imaginary cylinder about the axis of the _____.

(a) datum   (b) center   (c) feature   (d) outside diameter

_____   6.   It is important to select as datums surfaces that are _____.

(a) functional   (b) accurate   (c) internal   (d) external

_____   7.   Generally, concentricity is specified when coaxiality must be controlled independently of _____.

(a) alignment   (b) runout
(c) positional tolerance   (d) surface errors

_____   8.   According to General Rule 3, concentricity applies _____ unless otherwise specified.

(a) at LMC   (b) at MMC
(c) RFS   (d) at virtual condition

_____   9.   Since the form of the concentricity tolerance zone is a cylinder, the tolerance is specified as a _____.

(a) radius   (b) diameter   (c) total width   (d) radial width

_____   10.   A datum for concentricity may be a single cylinder, two separated cylinders used as one datum, or a (an) _____.

(a) axis   (b) center line
(c) flat face   (d) pair of lathe centers

CHAPTER 15 REVIEW PROBLEMS   PART II

In the part below, the smaller diameter is to be concentric RFS with the larger diameter within .002. Complete the drawing. (Hint: The shape of the tolerance zone is important.)

1.

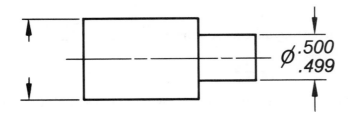

$\varnothing \dfrac{.500}{.499}$

In the part below, the ∅.998–1.000 is to be concentric RFS within .001 with the two small cylinders, used as one datum. Complete the drawing.

2.

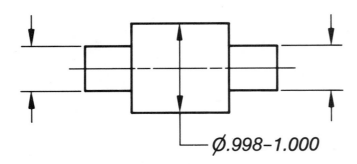

∅.998–1.000

The drawing below is an incomplete interpretation of the concentricity control in the part shown in Problem 2. Add an appropriate local note at each leader to complete the interpretation.

3.

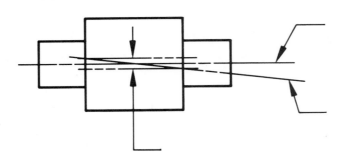

Complete the following statements:

4. Coaxiality can be controlled by any of three geometric characteristics. Concentricity is used when coaxiality must be _____ independently of _____ and when the tolerance must be held _____ _____.

5. Concentricity always applies RFS. When it is desirable to control coaxiality at MMC, _____ (another geometric characteristic) is specified.

CHAPTER 16 REVIEW PROBLEMS   PART I

Write the letter of the best answer in the space at left.

_____   1. Runout is any deviation of a surface from perfect form that
                   can be detected by _____.
                   (a) taking diametral measurements at 90°
                   (b) rotating the part about an axis
                   (c) a micrometer
                   (d) a comparator

_____   2. Runout is a _____ tolerance, including errors in round-
                   ness, straightness, perpendicularity, and coaxiality.
                   (a) composite (b) complex (c) multiple (d) positional

_____   3. In circular runout the deviation of each circular _____ is
                   controlled.
                   (a) feature (b) axis (c) surface (d) element

_____   4. Total runout controls the deviation of all _____ of a
                   surface, circular or straight.
                   (a) elements (b) features (c) datums (d) irregularities

_____   5. Circular runout is more commonly used because it is less
                   expensive to _____ and it is adequate for most design
                   functions.
                   (a) control (b) measure (c) specify (d) analyze

_____   6. The form of the tolerance zone for circular runout is the radial
                   distance between theoretical _____.
                   (a) cylinders (b) elements (c) circles (d) radii

_____   7. The form of the tolerance zone for total runout is the radial
                   distance between theoretical _____.
                   (a) cylinders (b) elements (c) circles (d) radii

_____   8. It is important to select as datums surfaces that are _____.
                   (a) functional (b) accurate (c) internal (d) external

_____   9. Where cylindrical parts are relatively large in diameter and
                   short in length, it is necessary to specify a (an) _____ as a
                   datum in addition to a cylindrical surface.
                   (a) axis (b) flat face (c) pair of lathe centers (d) element

_____   10. Runout tolerance always applies _____.
                   (a) at LMC (b) at MMC (c) RFS (d) at virtual condition

CHAPTER 16 REVIEW PROBLEMS   PART II

The circular runout of a certain diameter is to be held to .002 relative to one datum axis establishd by two diameters, A and B. Draw the appropriate feature control frame.

1.

The total runout of a diameter has to be maintained within .003 relative to another diameter and a perpendicular face. Draw an appropriate feature control frame. (Perpendicularity will not be specified.)

2.

If total runout tolerances of .002 are specified for Diameter A relative to Datum C and for Diameter B relative to Datum C, what is the resulting total runout tolerance between A and B? (Draw a sketch below, if you wish.)

3. _____

Complete the sentences below explaining how circular runout is inspected using a vee-block and dial indicator.

4. The _____ is placed in the vee-block, and

the part is rotated _____ degrees with a dial indicator held against the

_____ . The _____ of the dial indicator is equal to the

_____ error. Enough _____ are inspected

to ensure that all of them are within the specified _____.

(over)

What additional testing is required for total runout on a cylinder?

5. _____

_____

_____

In the stepped shaft shown below, the circular runout of the conical surface relative to the smaller cylinder must be held within .001. Complete the drawing.

6.

45° ± 2°

Under what two conditions is it best to use runout to control coaxiality, rather than concentricity or positional tolerance?

7.

_____

_____

_____

_____

CHAPTER 17 REVIEW PROBLEMS   PART I

Write the letter of the best answer in the space at left.

_____ 1. The true position of a feature is the _____ location of its axis or center plane from the feature or features from which it is dimensioned.

(a) theoretically exact   (b) approximate
(c) average   (d) accurate

_____ 2. Positional tolerance is the total _____ error in the location of a feature relative to another feature or to several other features.

(a) permissible (b) measurable (c) unilateral (d) bilateral

_____ 3. Positional tolerance is a _____ tolerance, including errors in roundness, straightness, parallelism, perpendicularity, and coaxiality.

(a) runout (b) multiple (c) complex (d) composite

The positional tolerance for cylindrical features such as holes and bosses is the ___(4)___ of the tolerance zone within which the ___(5)___ of the feature must lie.

_____ 4. (a) half-width (b) total width (c) radius (d) diameter

_____ 5. (a) center plane   (b) center line of symmetry
(c) axis   (d) boundary

_____ 6. The positional tolerance for noncylindrical features such as tabs and keyways is the _____ of the tolerance zone within which the center plane of the feature must lie.

(a) half-width (b) total width (c) radius (d) diameter

_____ 7. A .010 square coordinate tolerance zone has a diagonal permissible error of _____.

(a) .005 (b) .006 (c) .007 (d) .014

_____ 8. A coordinate tolerance zone is .016 square, with a diagonal tolerance of .022. The equivalent positional tolerance is .022 *diameter,* which allows _____ more permissible error than the .016 square tolerance zone.

(a) 25% (b) 50% (c) 57% (d) 75%

_____ 9. Location dimensions used with positional tolerance control are _____ dimensions.

(a) chain (b) basic (c) limit (d) toleranced

_____ 10. The axis of a hole located by a positional tolerance of ∅.014 can be out of perpendicular relative to the surface of the part by a maximum of _____.

(a) .0035   (b) .007   (c) .014   (d) .021

CHAPTER 17 REVIEW PROBLEMS   PART II

A part contains a number of holes that are in true position relative to two edges of
the part within ⌀.12 mm at MMC. Draw an appropriate feature control frame.

1.

The drawing below is the interpretation for a part with a large hole held in
positional tolerance relative to two edges. Complete the local notes.

2.

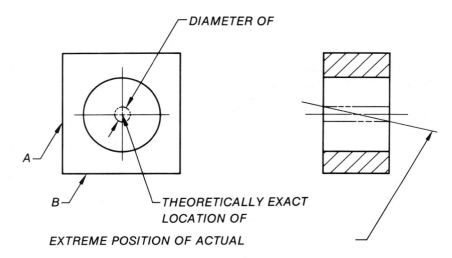

┌DIAMETER OF

A┐

B┘ └THEORETICALLY EXACT
LOCATION OF

EXTREME POSITION OF ACTUAL

In the drawing below, the two holes are in true position relative to the left and
bottom edges within ⌀.3 mm at MMC. Both datum surfaces are flat within
.15 mm. Complete the drawing, adding symbolism as necessary and location
dimensions for the holes.

3.

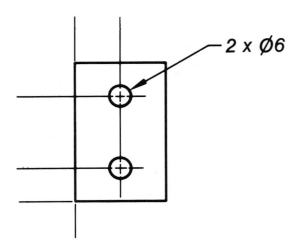

2 x ⌀6

A strap has four 5 mm ± .05 mm punched holes in line, with a positional tolerance of ∅.3 mm at MMC. Complete the tolerance variation table below for the possible produced sizes given.

4.

Produced Size of Hole	Positional Error Permissible
∅5.05 mm	_____
5.02	_____
5.00	_____
4.98	_____
4.97	_____

In the flange drawing below, Diameter A, a datum for the equally spaced holes, has its own positional tolerance (∅.003 Ⓜ). This is an application of Rule 5: A size feature that is used as a datum and is controlled by its own geometric tolerance applies at its *virtual* condition even though MMC is specified. Define virtual condition.

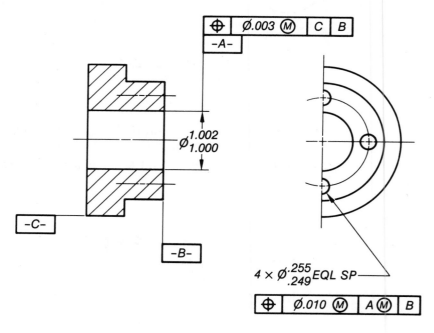

5. _____

_____

_____

What is the smallest acceptable diameter for Datum A?

6. _____

What is the permissible positional error of the equally spaced holes when they are produced at ∅.249 and Datum A is produced at ∅1.002?

7. _____

The drawing of the spacer below is an application of the control of an edge distance by specifying LMC.

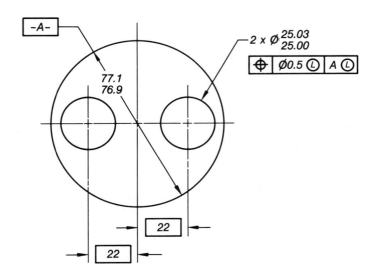

What is the edge distance when the holes and the datum are produced at their LMC sizes?

8. _____

What is the maximum positional tolerance for the two holes when they and the datum are *not* at LMC?

9. _____

If the two holes and the datum were produced at their MMC sizes, what would be the edge distance?

10. _____

CHAPTER 18 REVIEW PROBLEMS   PART I

Write the letter of the best answer in the space at left.

_____   1. In a floating fastener assembly, both parts have _____ holes.

(a) close-fitting (b) press-fit (c) clearance (d) tapped

_____   2. In a fixed fastener assembly, one of the mating parts has a _____ hole; the other, or others, have clearance holes.

(a) tight-fitting   (b) loose-fitting
(c) transition-fit   (d) lubricated

In the design of mating parts assembled with fasteners, it is necessary to make decisions about the amount of ___(3)___ required in the holes and of the ___(4)___ tolerance to be specified.

_____   3. (a) interference   (b) clearance
                 (c) perpendicularity (d) parallelism

_____   4. (a) size (b) positional (c) concentricity (d) projected

_____   5. For floating fasteners the positional tolerance is equal to the clearance times _____.

(a) ¼ (b) ½ (c) one (d) two

_____   6. For fixed fasteners the positional tolerance is equal to the clearance times _____.

(a) ¼ (b) ½ (c) one (d) two

_____   7. It sometimes occurs that a part may be rejected because a feature exceeds the size limits, even though it might actually fit the mating part. This can be prevented by specifying a _____.

(a) projected tolerance zone
(b) zero positional tolerance at MMC
(c) larger positional tolerance
(d) smaller positional tolerance

_____   8. Fitting a tab in a slot, using positional tolerance, is similar to a _____ situation.

(a) floating fasteners (b) fixed fasteners
(c) perpendicularity   (d) parallelism

_____   9. A special case of positional tolerance control wherein the shape of an object is the same but opposite on both sides of a center plane is called _____.

(a) coaxiality   (b) symmetry
(c) multiple patterns of features (d) composite tolerance

_____ 10. Most patterns of features located by basic dimensions from the same datums are considered one _____ pattern.

(a) complex   (b) compound

(c) composite (d) complicated

CHAPTER 18 REVIEW PROBLEMS   PART II

Write a local note to replace the feature control frame in the drawing below.

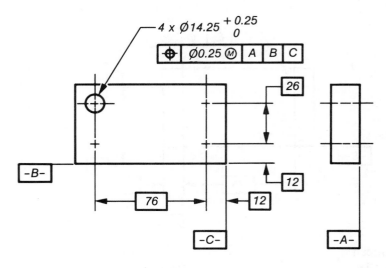

1. _____

_____

_____

To the drawing below, add a requirement that the axis of the hole be perpendicular to the top surface within ⌀.3 mm at MMC and a requirement that the perpendicularity tolerance be projected 14 mm above the top surface.

2.

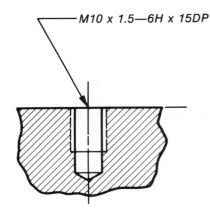

The next drawing is an application of zero positional tolerance at MMC. The only variation that is allowed is in the size of the hole. Complete the tolerance variation table below for the given produced sizes of the hole.

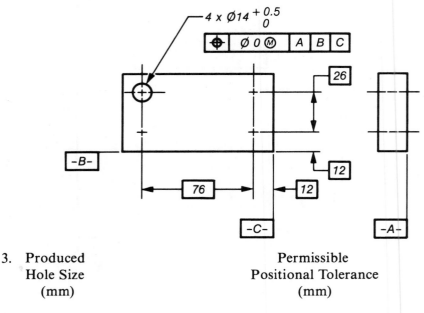

3.

Produced Hole Size (mm)	Permissible Positional Tolerance (mm)
⌀14	_____
⌀14.2	_____
⌀14.5	_____

In the situation shown in Problem 3, 13.9–14.0 mm screws are to be inserted in the four holes. What will be the least clearance (tightest fit) between the screws and the holes when:

4. (a) the holes are produced at MMC? _____

    (b) the holes are produced at ⌀14.4 mm? _____

    (c) What is the maximum possible clearance (loosest fit) between the screws and holes? _____

The formula used to determine the clearance and positional tolerance in the design of mating tabs and slots is similar to the floating fastener formula. Write the formula below, using

T_{tot} for Total tolerance in both parts

W_s for Width of slot at MMC

W_t for Width of tab at MMC

5. _____

The drawing below shows a disk with two keyways located by positional tolerance.

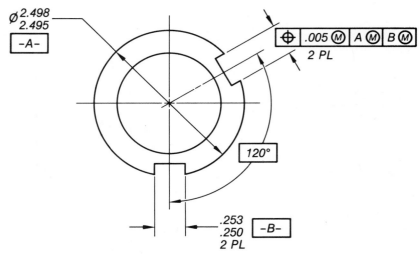

6. (a) What geometric characteristic (besides positional tolerance) is actually being controlled?

_____

(b) Write a local note that might replace the feature control frame.

_____

_____

_____

(c) Give one possible reason why the outside diameter is the more important of the two datums.

_____

For the drawing in Problem 6, complete the tolerance variation table below for the possible produced sizes shown. (Hint: the total departure from **MMC** for all the sizes involved must be found.)

7.

| Produced Size | | | Permissible |
Upper Slot	Lower Slot	O.D.	Positional Tolerance
.253	.253	2.498	_____
.252	.252	2.497	_____
.251	.251	2.496	_____
.250	.250	2.495	_____

The following questions refer to the drawing below of a plate with multiple patterns of features (two hole-patterns).

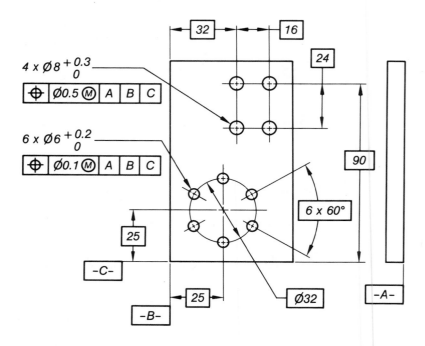

8. (a) Should the hole patterns be inspected as (check one) _____ separate patterns? _____ one composite pattern?

(b) Give the rule concerning multiple patterns of features that supports your answer to Problem 8 (a).

_____

_____

_____

(c) How many gages will be used to inspect the location of the holes? _____

Say it is desired to treat the two hole-patterns in Problem 8 not as one composite pattern but as separate requirements. Redraw the two feature control frames in the space below, revising them if necessary.

9.

Write notes to interpret the composite feature control frame for the circular hole-pattern in the drawing below.

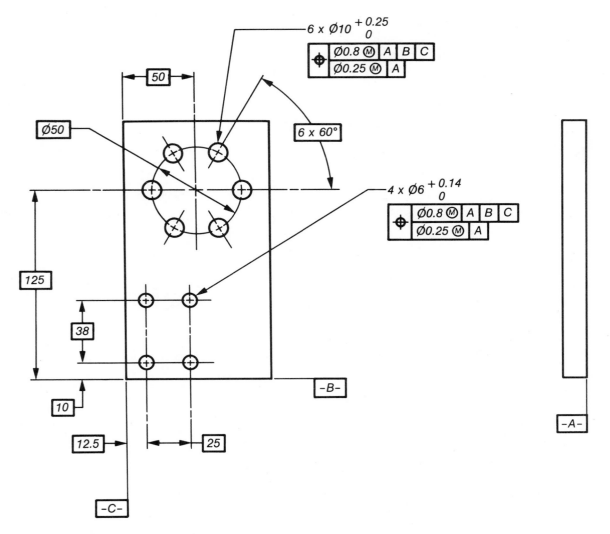

6 x Ø10 $^{+0.25}_{\ \ 0}$

⌖	Ø0.8 Ⓜ	A	B	C
	Ø0.25 Ⓜ	A		

4 x Ø6 $^{+0.14}_{\ \ 0}$

⌖	Ø0.8 Ⓜ	A	B	C
	Ø0.25 Ⓜ	A		

Ø50

6 x 60°

50

125

38

10

12.5   25

-B-

-A-

-C-

10. (a) The pattern location _____

_____

_____

     (b) The individual holes _____

_____

_____

CHAPTER 19 REVIEW PROBLEMS   PART I

Write the letter of the best answer in the space at left.

_____   1. Two or more geometrical figures are coaxial if they have the
               same _____.
               (a) center   (b) center plane   (c) axis   (d) datum

_____   2. There is no symbol for the general case of coaxiality, but there
               are _____ symbols denoting special cases.
               (a) two   (b) three   (c) four   (d) five

_____   3. The form of the tolerance zone for coaxial positional tolerance
               is an imaginary _____ perfectly coaxial with the datum
               axis.
               (a) circle   (b) cylinder   (c) ring   (d) tube

_____   4. The tolerance for coaxial positional tolerance is always
               expressed as a _____.
               (a) radius   (b) diameter   (c) half-width   (d) total width

_____   5. Coaxial positional tolerance is a _____ tolerance because
               it includes surface errors.
               (a) composite   (b) complex
               (c) complicated   (d) compound

_____   6. When the reason for specifying coaxiality is to ensure that
               parts will assemble properly, the modifier used is _____.
               (a) LMC   (b) MMC   (c) RFS   (d) virtual condition

_____   7. The most common application of the LMC modifier is to
               control a wall thickness or _____.
               (a) edge distance   (b) basic dimension
               (c) location dimension (d) assemblability

_____   8. In using positional tolerance to control coaxial holes, the
               location tolerance for the holes as a group requires one or
               more datums. The coaxiality tolerance for the individual holes
               requires _____.
               (a) no datum   (b) one datum
               (c) two datums   (d) one or more datums

_____ 9. A stepped shaft (two diameters) is required to assemble into a stepped sleeve. In calculating the fit, the total positional tolerance for both parts is equal to the _____.

(a) sum of the maximum clearances
(b) sum of the minimum clearances
(c) difference between the larger diameters
(d) difference between the smaller diameters

_____ 10. Zero positional tolerancing is a practical method of controlling coaxiality while avoiding _____.

(a) unused positional tolerance (b) excess clearance
(c) insufficient clearance   (d) excess runout

CHAPTER 19 REVIEW PROBLEMS   PART II

In the seal body drawn below, the small diameter must be coaxial with the large diameter within ⌀.001 when both diameters are at MMC. Complete the drawing.

1.

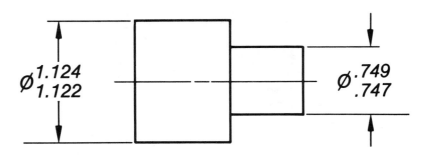

$\varnothing{}^{1.124}_{1.122}$ $\varnothing{}^{.749}_{.747}$

The outside diameter of a thin-wall sleeve must be in true position within ⌀.1 mm at LMC relative to the inside diameter at LMC. Complete the drawing.

2.

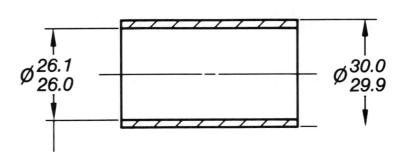

$\varnothing{}^{26.1}_{26.0}$ $\varnothing{}^{30.0}_{29.9}$

The small diameter of the balance block shown below must be coaxial within ⌀.0005 at MMC with the large diameter no matter what the produced size of the large diameter might be. Also, the large diameter at MMC must be in true position within ⌀.010 relative to the height and width of the block at their MMC sizes. Complete the drawing.

3.

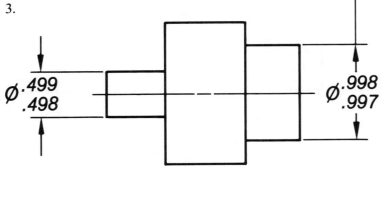
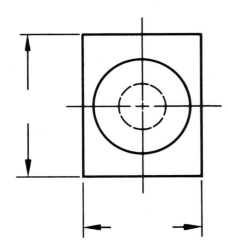

$\varnothing{}^{.499}_{.498}$ $\varnothing{}^{.998}_{.997}$

(over)

In the hinge body below, the four ∅10 mm-holes are to be coaxial within ∅.1 mm at MMC and in true position within ∅.2 mm at MMC relative to two perpendicular flat surfaces. Add a composite feature control frame and datum feature symbols to express this design intent.

4.

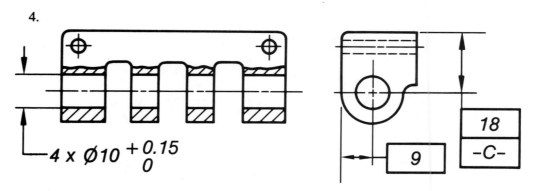

$$4 \times \varnothing 10 \, {}^{+0.15}_{\ \ 0}$$

Below are two drawings of a socket and a mating plug. Use the three-step procedure to calculate the positional tolerance for each part, making the tolerance for the socket larger than the plug tolerance. Add the positional tolerances to the drawings.

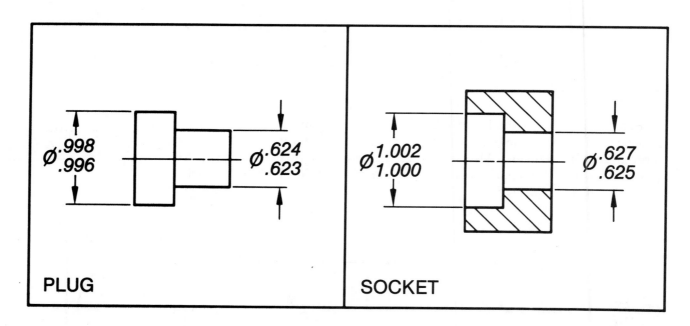

5. Step 1   Obtain the minimum clearances.

    Small diameters: _____   Large diameters: _____

    Step 2   Add the minimum clearances to obtain the total positional tolerance for both parts.

    The sum is _____.

    Step 3   Divide the total positional tolerance between the two parts.

    Tolerance on socket: _____

    Tolerance on plug: _____

Do not forget to add the positional tolerances to the drawings.

The same seal body as in Problem 1 is drawn below but with a zero positional tolerance at MMC controlling coaxiality of the two diameters. Make up a tolerance variation table for two possible produced sizes: the MMC and the LMC for both diameters. Show the total departure (if any) from MMC resulting from the produced sizes. Show also the resulting permissible positional error. Neatly letter appropriate headings.

6.

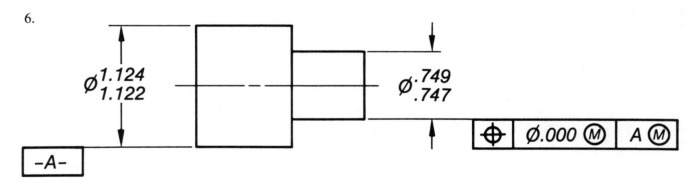

CHAPTER 20 REVIEW PROBLEMS   PART I

Write the letter of the best answer in the space at left.

_____   1. On some parts it is necessary to specify the desired contact
                    points between the datum feature and the datum plane (tool-
                    ing). This must be done in _____ planes.

                    (a) two   (b) three   (c) five   (d) six

_____   2. An unsupported object in space can be moved in _____
                    directions.

                    (a) two   (b) three   (c) five   (d) six

_____   3. The specified points where contact is made with the object are
                    called _____.

                    (a) datum targets or tooling points
                    (b) datum features
                    (c) supporting points
                    (d) tooling targets

_____   4. A system of three datum planes at right angles is called a
                    _____.

                    (a) tooling setup   (b) datum frame
                    (c) position fix   (d) manufacturing frame

_____   5. After the first and second datums are selected, a remaining
                    _____ feature is used as the third datum.

                    (a) accurate   (b) flat
                    (c) mutually accessible   (d) mutually perpendicular

_____   6. Generally, the maximum number of points at which contact is
                    made with supporting tooling is _____.

                    (a) six   (b) eight   (c) ten   (d) twelve

_____   7. The datum identifying letter (A, B, or C) followed by a target
                    number is drawn _____ the datum target symbol.

                    (a) in the lower half of   (b) in the upper half of
                    (c) above   (d) below

_____   8. Datum targets, although often referred to as tooling *points*,
                    may be in any of _____ forms.

                    (a) two   (b) three   (c) five   (d) six

_____   9. When the datum target is a line in contact with the part, it is
                    represented on the drawing as a (an) _____ line.

                    (a) extension   (b) center   (c) broken   (d) phantom

_____ 10. A cylindrical surface has a location ability equivalent to _____ datum planes.

(a) two (b) three (c) five (d) six

_____ 11. In a datum frame for a cylindrical part, the part is supported on the cylindrical surface and _____.

(a) at a tooling hole (b) along the axis
(c) on lathe centers (d) at one flat face

_____ 12. When an entire cylinder is used as a datum and no datum targets are specified, the diameter of the supporting tooling depends on whether the cylinder is an inside or outside diameter and the _____ specified.

(a) modifier (b) geometric tolerance
(c) basic dimensions (d) local notes

_____ 13. When datum targets are specified on revolving shafts, the datum target locations duplicate the positions of the _____.

(a) bearings (b) lathe centers
(c) flat faces (d) grinding reliefs

_____ 14. On drawings with a step datum, the amount of offset is dimensioned for the supporting tooling with a basic dimension. The offset on the part itself may be given by a _____.

(a) reference dimension (b) positional tolerance
(c) parallelism tolerance (d) size dimension

_____ 15. Equalizing pins (tooling) are used to locate _____.

(a) round-end parts (b) step datums
(c) bosses (d) vee-shaped parts

CHAPTER 20 REVIEW PROBLEMS   PART II

On the isometric drawing below of an unsupported object in space, select suitable datum targets (tooling points) on the three visible surfaces to make a complete datum frame. Show the point locations with crosses. Omit dimensions.

1.

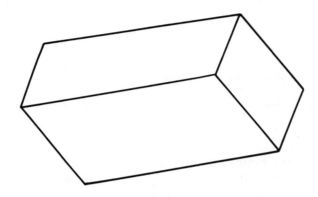

Add datum target symbols in both views to the drawing below, complete with datum identifying letters and target numbers.

2.

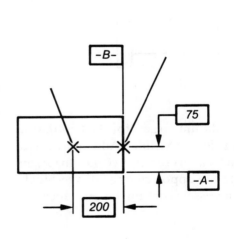

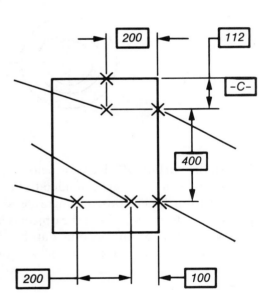

In the unfinished drawing below, the two small cylinders together will be Datum A, and the right end face of the shaft will be Datum B. Datum A will be identified by a target line on the left and a target area on the right. The large diameter will be coaxial within ∅.001 with the two small cylinders together regardless of the size of the features. Complete the drawing to show the design intent. Omit shaft size dimensions.

3.

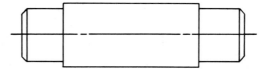

Add a complete datum frame to the cast tube below. Select suitable datum targets on cast surfaces. Label the datum planes, give location dimensions for all datum targets, and show all datum target symbols.

4.

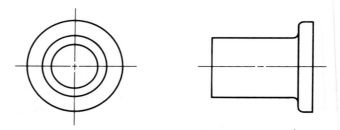

In the partial full-size view below, two of three datum targets for the first datum are shown. They are intended to be on the far side (not visible). Add appropriate location dimensions, datum target symbols, and leaders for the two datum targets only. The round end of the object is intended to be supported by a 90° vee-block, which will be the second datum. Add the necessary datum targets, location dimensions, and datum target symbols.

5.

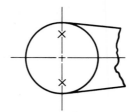

COMPREHENSIVE EXERCISE 1

Definitions and Symbols

In the parentheses in the right-hand column, place the number of the matching term from the left-hand column. Use every term in the left-hand column.

1. Angularity
2. Basic Dimension
3. Bilateral Tolerance
4. Concentricity
5. Cylindricity
6. Datum
7. Flatness
8. Limits
9. Maximum Material Condition
10. Parallelism
11. Perpendicularity
12. Positional Tolerance
13. Profile of a Line
14. Profile of Surface
15. Regardless of Feature Size
16. Roundness
17. Runout, circular
18. Runout, total
19. Straightness
20. Tolerance
21. Unilateral Tolerance

( ) Where the tolerance of form or position must be met regardless of where the feature lies within its size tolerance.

( ) //

( ) ⌀

( ) The theoretical value used to describe the exact size, shape, or location of a feature.

( ) ▱

( ) ∠

( ) One in which variation is permitted in both directions from the specified dimension.

( ) One in which variation is permitted in only one direction from the specified dimension.

( ) ⌒

( ) ⟋⟋

( ) ◎

( ) ⟋

( ) ⊕

( ) The condition of a part feature when it contains the maximum amount of material.

( ) ⊥

( ) ──

( ) The total amount by which a dimension may vary.

( ) ○

( ) ⌒

( ) The maximum and minimum sizes indicated by a toleranced dimension.

( ) A surface indicated on the drawing, from which measurements may be made.

COMPREHENSIVE EXERCISE 2

Applications of Geometric Tolerances

Add feature control frames for flatness, roundness, cylindricity, parallelism, perpendicularity, runout, concentricity, and positional tolerance. Select appropriate datums where needed.

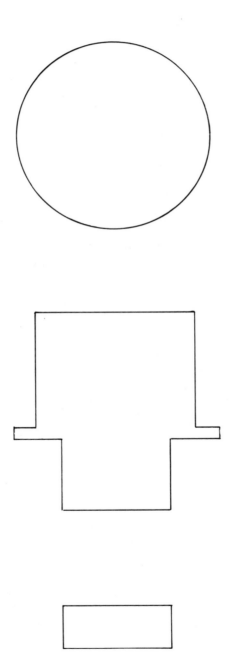

COMPREHENSIVE EXERCISE 3

Applications of Geometric Tolerances

Add feature control frames and datum identifying symbols to the drawing on the following page to specify the following requirements. All symbols and frames are to be about the correct size.

Surface 1 to be round within .0005.

Surface 5 to be flat within .0005.

Surface 2 to be cylindrical within .001.

Surfaces 2, 3, and 4 to be concentric with surface 1 within ⌀.002.

Surfaces 5 and 6 to be parallel within .003.

Runout of surface 7 when the part is supported on surfaces 1 and 5 to be .003, maximum.

For each feature control frame and datum symbol, write an appropriate local note which might be used in place of the symbology. List them below by the numbers corresponding to numbers shown on the drawing.

1. This surface to be round within .0005. _____

2. _____

3. _____

    _____

4. _____

    _____

5. _____

    _____

6. _____

    _____

7. _____

    _____

(over)

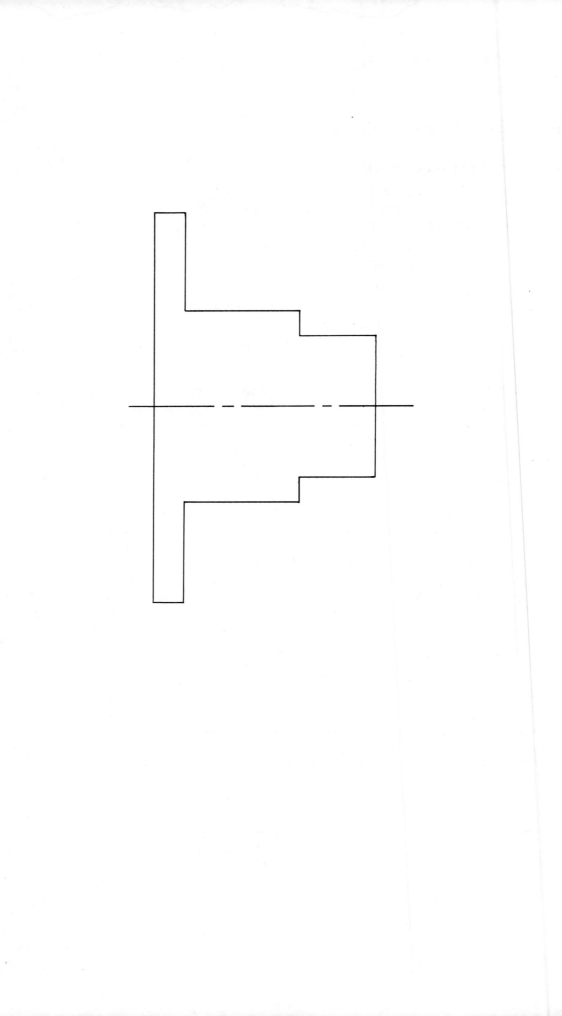

COMPREHENSIVE EXERCISE 4

Calculations for Fasteners

A lid assembles onto a gear case with six ⁵/₁₆–18 cap screws. Using a positional tolerance of ∅.015 at MMC for the screw hole locations in both parts, calculate the size of the clearance holes in the lid, selecting a drill size from the standard sizes listed at right. Perform all calculations neatly, starting with a formula.

1. _____

Drill Sizes

.312
.316
.323
.328
.332
.339
.344
.348
.358

Three parts are held together with ½–12 bolts. The clearance holes in all three parts are ∅.531. What positional tolerance at MMC will be required in the drawings of the three parts? Perform all calculations neatly, starting with a formula.

2. _____

COMPREHENSIVE EXERCISE 5

Calculations for Slots and Tabs

All the questions in this exercise pertain to the two drawings of mating parts, below.

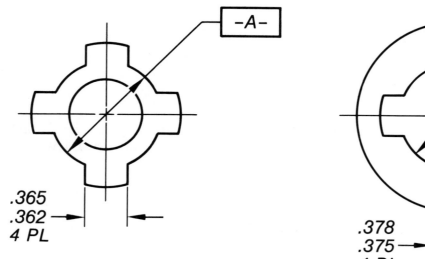

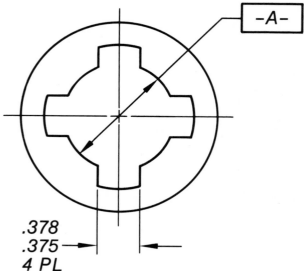

.365
.362
4 PL

.378
.375
4 PL

Calculate the positional tolerance at MMC for the tabs and the slots, and add the appropriate feature control frames. Make the tab tolerance two-thirds of the slot tolerance. Show all calculations, starting with a formula.

1. Tab tolerance: _____   Slot tolerance: _____

(over)

What is the maximum permissible positional error when the tabs are at their LMC sizes? Show all calculations.

2. _____

What would be the maximum permissible error when the tabs are at their LMC sizes if the positional tolerance were specified RFS? Show all calculations.

3. _____

COMPREHENSIVE EXERCISE 6

Calculations for Fit of Coaxial Parts

You have been asked to design the fit of the socket and mating plug shown in the two drawings below.

Given:

The internal diameters of the socket:

See drawing

The tightest fit with permissible out-of-coaxiality:

Metal-to-metal

The loosest fit with perfect coaxiality:

.006

The coaxiality tolerances:

Plug $\varnothing$.001
Socket $\varnothing$.002

Required:

Decide whether the coaxiality tolerances and the datums should apply at MMC or RFS.

Calculate limit dimensions for the plug diameters. (There is more than one correct answer.) Calculations may be done on scratch paper.

Add the missing information to both drawings.

You are not concerned with dimensions not related to the fit. A drafter will place those on the formal drawings later.

(Hint: In the three-step procedure for calculating coaxial fits, it was said that the total tolerance equals the sum of the clearances.)

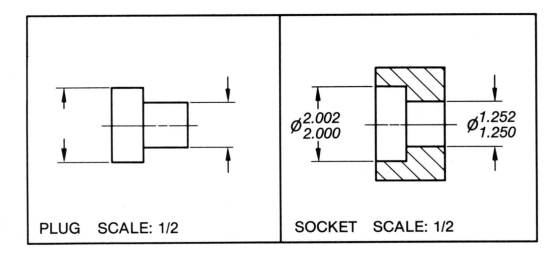

| PLUG   SCALE: 1/2 | SOCKET   SCALE: 1/2 |

COMPREHENSIVE EXERCISE 7

Conversion—Coordinate Tolerances to Positional Tolerances

The answers to all the problems in this exercise can be found by using the conversion tables in Appendix B. Round off to the nearest thousandth unless otherwise specified.

The following bilateral coordinate tolerances are given on a drawing in two perpendicular directions (square tolerance zone). Convert them to equivalent round positional tolerance values.

1. $\pm.010 = \varnothing$ _____   2. $\pm.0025 =$ _____

3. $\pm.008 =$ _____   4. $\pm.005 =$ _____

5. $\pm.016 =$ _____   6. $\pm.0005 =$ _____

Convert the following square tolerance zone values to equivalent round positional tolerance zone sizes.

7. $.015 =$ _____   8. $.008 =$ _____

9. $.020 =$ _____   10. $.03 =$ _____
 (.xx)

11. $.002 =$ _____   12. $.0005 =$ _____
 (.xxxx)

Convert the following rectangular tolerance zone values to equivalent round positional tolerance zone sizes.

13. $.005 \times .010 =$ _____   14. $.008 \times .012 =$ _____

15. $.003 \times .004 =$ _____   16. $.006 \times .009 =$ _____

In the inspection of a baseplate, a coordinate inspection machine was used to measure deviation from basic hole location dimensions from two perpendicular datums (X and Y), and the data given below were recorded. Convert each pair of values to equivalent round positional tolerance values, rounding off to the nearest thousandth.

Hole No.	Deviation X	Y	Positional Tolerance
17.	.002	.004	_____
18.	.008	.006	_____
19.	.005	.001	_____
20.	.007	.003	_____

A large number of $5/16$-diameter holes are drilled in a housing. The print shows that the drill tolerance is +.006 –.002 and the holes must be in true position at MMC within .015 diameter. The deviations of the holes from their basic location dimensions are measured and recorded. The actual hole diameters are also measured and recorded. The table below presents the recorded data and provides spaces for other derived (figured out) data to be filled in by the inspector (the student). For each hole, obtain the equivalent positional error at MMC, the *maximum* permissible positional error, and indicate by a check mark whether the hole location is acceptable or not. (Hint: Remember that when the hole as produced is not at its MMC, a bonus tolerance is available.)

Hole No.	Deviation X	Y	Equivalent Positional Error	Actual Hole Size	Maximum Permissible Positional Error	Acceptable	Not Acceptable
21.	.006	.005	.016	.310	.015	_____	✓
22.	.007	.004	_____	.312	_____	_____	_____
23.	.002	.003	_____	.315	_____	_____	_____
24.	.008	.006	_____	.318	_____	_____	_____
25.	.007	.008	_____	.314	_____	_____	_____